# POPULAR FICTION AND BRAIN SCIENCE IN THE LATE NINETEENTH CENTURY

In the 1860s and 1870s, leading neurologists used animal experimentation to establish that discrete sections of the brain regulate specific mental and physical functions. These discoveries had immediate medical benefits: David Ferrier's detailed cortical maps, for example, saved lives by helping surgeons locate brain tumors and hemorrhages without first opening up the skull. These experiments both incited controversy and stimulated creative thought, because they challenged the possibility of an extra-corporeal soul. This book examines the cultural impact of neurological experiments on late-Victorian Gothic romances by Robert Louis Stevenson, Bram Stoker, H. G. Wells, and others. Novels like *Dracula* and *Dr. Jekyll and Mr. Hyde* expressed the deep-seated fears and visionary possibilities suggested by cerebral localization research, and offered a corrective to the linearity and objectivity of late-Victorian neurology.

ANNE STILES is Assistant Professor of English at Saint Louis University.

CAMBRIDGE STUDIES IN NINETEENTH-CENTURY LITERATURE AND CULTURE



Nineteenth-century British literature and culture have been rich fields for interdisciplinary studies. Since the turn of the twentieth century, scholars and critics have tracked the intersections and tensions between Victorian literature and the visual arts, politics, social organization, economic life, technical innovations, scientific thought – in short, culture in its broadest sense. In recent years, theoretical challenges and historiographical shifts have unsettled the assumptions of previous scholarly synthesis and called into question the terms of older debates. Whereas the tendency in much past literary critical interpretation was to use the metaphor of culture as "background," feminist, Foucauldian, and other analyses have employed more dynamic models that raise questions of power and of circulation. Such developments have reanimated the field. This series aims to accommodate and promote the most interesting work being undertaken on the frontiers of the field of nineteenth-century literary studies: work which intersects fruitfully with other fields of study such as history, or literary theory, or the history of science. Comparative as well as interdisciplinary approaches are welcomed.

*A complete list of titles published will be found at the end of the book*

# POPULAR FICTION AND BRAIN SCIENCE IN THE LATE NINETEENTH CENTURY

ANNE STILES

CAMBRIDGE
UNIVERSITY PRESS

CAMBRIDGE UNIVERSITY PRESS
Cambridge, New York, Melbourne, Madrid, Cape Town,
Singapore, São Paulo, Delhi, Tokyo, Mexico City

Cambridge University Press
The Edinburgh Building, Cambridge CB2 8RU, UK

Published in the United States of America by Cambridge University Press, New York

www.cambridge.org
Information on this title: www.cambridge.org/9781107010017



First published 2012

Printed in the United Kingdom at the University Press, Cambridge

*A catalogue record for this publication is available from the British Library*

*Library of Congress Cataloguing in Publication data*
Stiles, Anne 1975–
Popular fiction and brain science in the late nineteenth century / Anne Stiles.
p. cm. – (Cambridge studies in nineteenth-century literature and culture; 78)
Includes bibliographical references and index.
ISBN 978-1-107-01001-7
1. Gothic fiction (Literary genre), English–History and criticism. 2. English fiction–19th century–History and criticism. 3. Literature and science–Great Britain–History–19th century. 4. Literature and medicine–Great Britain–History–19th century. 5. Mind and body in literature. 6. Physiology in literature. 7. Gothic revival (Literature)–Great Britain–History–19th century. 8. Neurosciences–Great Britain–History–19th century. I. Title.
PR878.T3S75 2012
823′.0872909–dc23
2011038185

ISBN 978-1-107-01001-7 Hardback

*For Kevin*

# *Contents*

*List of illustrations* *page* viii
*Acknowledgments* ix

Introduction: cerebral localization and the late-Victorian Gothic romance 1

PART I REACTIONARIES

1 Robert Louis Stevenson's *Jekyll and Hyde* and the double brain 27

2 Bram Stoker's *Dracula* and cerebral automatism 50

PART II MATERIALISTS

3 Photographic memory in the works of Grant Allen 85

PART III VISIONARIES

4 H. G. Wells and the evolution of the mad scientist 119

5 Marie Corelli and the neuron 156

Epilogue: looking forward 185

*Notes* 191
*Bibliography* 232
*Index* 248

# *Illustrations*


1 Sir David Ferrier, "The Brain of a Criminal Lunatic," *Brain: A Journal of Neurology* 5 (April 1882): 62–73, 67 (courtesy of Oxford University Press) *page* 39

2 Sir William Thornley Stoker, "Memorandum on Head Injuries," *c.*1890–6 (courtesy of the Rosenbach Museum and Library) 63

3 Ferrier, "Upper Surface of the Hemispheres of the Monkey," from Sir David Ferrier, *Functions of the Brain* (New York: G. P. Putnam's Sons, 1876), 156 (reprinted courtesy of Penguin Publishing) 66

4 Eadweard Muybridge, photograph of a trotting horse, 1878 (courtesy of the US Library of Congress) 109

5 "1,000,000 A.D.," *Punch, or the London Charivari* (November 25, 1893): 250 120

6 Jacques Moreau, "Tree of Nervosity," in *La psychologie morbide* (Paris: Librairie Victor Masson, 1859) 129

# *Acknowledgments*

This book began as a UCLA dissertation that was partially funded by a University of California Humanities Research Institute Andrew V. White Scholarship. My advisor, Joseph Bristow, provided generous and insightful mentoring during the early stages of this project. He also offered crucial advice and encouragement as I strove to reshape the dissertation into the present monograph. I am grateful for his continued support, and for that of my other committee members, Jonathan Grossman and Anne Mellor, whose feedback on several drafts of the dissertation strengthened the project immensely.

The American Academy of Arts and Sciences generously provided support for research and writing during the 2006–7 academic year, including office space and access to Harvard University Libraries. I am grateful to my colleagues at the Academy for providing helpful advice on drafts of Chapters 4 and 5. I am also thankful to Anne Harrington for encouraging me to audit two of her courses on mind–body medicine, both of which forever changed the way I think about the history of science. I also learned much from the Seminar on Cognition and the Arts at the Harvard Humanities Center. Alan Richardson kindly welcomed me into the group and invited me to speak on one occasion. He also provided valuable perspective on the relationship between cerebral localization and phrenology that I touch upon in the introduction.

The later stages of this project were supported by a Barbara Thom Postdoctoral Fellowship at the Huntington Library in San Marino, California. My colleagues at the Huntington provided much-needed advice and moral support, particularly Paul Gilmore, Catherine Molineaux, Adam Shapiro, Mary Terrall, and Susan Gillman. Thanks are due to the English Department at Washington State University, where I worked from August 2007 to May 2011, for permission to take two research leaves in four years in order to complete this book. Department

Chair George Kennedy was unflagging in his support of my research, as were my faculty mentors at WSU, Donna Campbell and Debbie Lee. All three offered expert feedback and encouragement at crucial points during the writing process. WSU librarian Bob Matuozzi, meanwhile, provided timely help in locating images for Chapters 1 and 4. I am also grateful to my current employer, the Department of English at Saint Louis University, for continued support of my research into the intersections between neurology and literature.

This book has benefited from the advice and support of colleagues from the North American Society for Victorian Studies and the International Society for the History of the Neurosciences. Nicholas Dames, Kate Flint, and Daniel Novak kindly invited me to speak on panels at various NAVSA conferences, where I met like-minded individuals pursuing the intersections between literature and medicine. Stan Finger and his colleagues at ISHN supported my work while it was still in the very early stages, and invited me to speak on the connections between vampires and neuroscience. Russell Johnson deserves thanks for introducing me to ISHN and helping me find my way around UCLA's Louise M. Darling Biomedical Library.

Other colleagues who offered advice, references, and feedback on various chapter drafts include Paul Brians, Mark S. Micale, Andrew Scull, Laura Otis, Jill Galvan, Patricia Meyer Spacks, Neil Hultgren, Alison Harvey, and Heather Wozniak, each of whom contributed valuable insights to this project. Julia Reid and Peter Morton, meanwhile, generously shared with me unpublished manuscript versions of their work. I am also grateful for the support of Linda Bree at every stage of the editorial process, and for the advice of the two anonymous readers, who helped transform the manuscript into a more cohesive and engaging document.

Earlier versions of three chapters have been published previously. Chapter 1 is reprinted with permission from *SEL: Studies in English Literature 1500–1900* 46.4 (Autumn 2006), while a portion of Chapter 2 appeared in *Journal of the History of the Neurosciences* 15 (2006). A shortened version of Chapter 4, entitled "Literature in *Mind*: H. G. Wells and the Evolution of the Mad Scientist," was originally published © 2009 *Journal of the History of Ideas* 70.2 (April 2009). My thanks go to all three journals for permission to reprint material. I also wish to thank the Rosenbach Museum and Library for permission to publish Sir William Thornley Stoker's "Memorandum on Head Injuries" in Chapter 2, and Oxford University Press for permission to reprint David Ferrier's illustration of "The Brain of a Criminal Lunatic" in Chapter 1. The cover image,

Paul Cézanne's "Still Life with Skull" (1865/7), is reproduced courtesy of the Kunsthaus Zürich.

On a personal level, this project owes much to my parents, David and Cindy Stiles, for imparting to me their love of history, literature, and fine art. Thanks also to my flock, whose cheerful chirps always brighten up the writing process. But most of all, I wish to acknowledge my husband, Kevin Kelsey, who has criss-crossed the country with me and this project, and who has read every draft along the way. It is to him that I dedicate this book.

# *Introduction: cerebral localization and the late-Victorian Gothic romance*

In Bram Stoker's *Dracula* (1897), madhouse superintendent Dr. John Seward lauds the work of two prominent physiologists who advanced the study of neuroscience by experimenting on live animals:

> Men sneered at vivisection, and yet look at its results today! Why not advance science in its most difficult and vital aspect – the knowledge of the brain? Had I even the secret of one such mind – did I hold the key to the fancy of even one lunatic – I might advance my own branch of science to a pitch compared with which Burdon-Sanderson's physiology or Ferrier's brain-knowledge would be as nothing.[1]

Critics have tended to overlook this passage, and perhaps understandably so – on the surface, neurological experiments seemingly have little to do with vampires, crucifixes, and the other supernatural mysteries at the heart of Stoker's novel. But in fact, a series of neurological experiments that began in the 1860s and 1870s – conducted by Sir David Ferrier, among others – had a profound impact on late-Victorian Gothic novels and romances such as *Dracula*. In turn, these novels often influenced the direction of future neurological research. This seemingly unlikely, symbiotic relationship between *fin-de-siècle* neurology and certain kinds of popular fiction extends to matters of form as well as content. In this study, I show how late-Victorian neurologists and authors of Gothic romances shared a fascination for boundaries and their transgression, especially the evanescent mind–body divide and the limits of human free will. These shared philosophical concerns help to explain the surprising number of brains, brain cells, and neurological references in late-Victorian Gothic novels and romances. At the same time, novelists did not simply accept a neurological perspective. Instead, through their snarled plotlines and depictions of tormented subjectivity, Gothic romances often criticized the objective, linear viewpoint of late-Victorian neurological science, not to mention its sometimes rigid biological determinism.

The late-Victorian and Edwardian periods witnessed watershed developments in neurological science, particularly the cerebral localization experiments of scientists such as David Ferrier and John Hughlings Jackson in England, Paul Broca in France, and Gustav Fritsch and Eduard Hitzig in Germany. These experiments began in earnest in 1861, when Broca linked the third frontal convolution of the left brain hemisphere to linguistic ability. Broca had found that speech difficulties such as aphasia frequently occurred after damage to this particular portion of the brain.[2] Following Broca's lead, other scientists attempted to establish that discrete sections of the brain regulate specific mental and physical functions. In the late 1860s, for instance, Jackson traced certain defects of memory to lesions of the posterior part of the right hemisphere, hypothesizing that the right hemisphere was involved in spatial perception.[3]

Jackson and Broca had relied primarily on autopsies and clinical case studies to arrive at their discoveries. By contrast, during the 1870s, scientists performed experimental surgeries on a variety of laboratory animals in an effort to localize cerebral functions more precisely. For example, Prussian scientists Fritsch and Hitzig operated on numerous dogs (sometimes without anesthesia) in order to determine that muscular movements were controlled by the motor cortex of the brain.[4] In the process, they also proved that the substance of the brain was electrically excitable, a possibility denied by previous scientists.[5] Building on these experiments, Scottish physician Ferrier exposed and electrically stimulated the brains of cats, dogs, rabbits, and monkeys. By exciting various regions of these animal brains and observing the results, then extrapolating these results to corresponding areas of the human cerebral cortex, Ferrier produced elegant cortical maps detailing which areas of the brain controlled specific functions.

The medical benefits of these discoveries became apparent almost immediately. For instance, Ferrier's detailed cortical maps saved lives by helping surgeons locate brain tumors and hemorrhages without first opening up the skull.[6] Cerebral localization experiments incited controversy, however, because they challenged the possibility of free will or an extra-corporeal soul. Indeed, ever since René Descartes contended that the soul resides in the pineal gland of the brain, attempts to correlate specific regions of the brain with particular functions have raised theological hackles. Late-Victorian debates surrounding cerebral localization resonated far beyond the professional scientific community, infiltrating the popular press and popular literature.

*Popular Fiction and Brain Science in the Late Nineteenth Century* explores the seemingly paradoxical fact that British popular novelists – those associated with commercially successful genres such as the romance, the Gothic novel, and the "shilling shocker" – were often exceptionally well informed about neurological theories and their philosophical ramifications, more so than many respected practitioners of realism. This book examines the works of scientifically savvy popular novelists including Robert Louis Stevenson, Bram Stoker, H. G. Wells, Grant Allen, and Marie Corelli, some of the most financially successful and culturally influential authors of their time. Their fictions collectively demonstrate how popular developments like the late-Victorian romance revival and the simultaneous revival of the Gothic expressed the deep-seated fears spawned by cerebral localization. Furthermore, these novels shaped scientific thought and influenced public opinion toward neurological innovations.

The question this volume addresses is why certain novelists proved more responsive to (though hardly uncritical of) neurological theories than others. Specifically, why did Gothic romances like Robert Louis Stevenson's *Strange Case of Dr. Jekyll and Mr. Hyde* (1886), Bram Stoker's *Dracula* (1897), and H. G. Wells's *The War of the Worlds* (1898) probe the implications of cerebral localization experiments more often and more deeply than "serious" literary genres like the realist novel? In the course of addressing this query, I hope to problematize the frequently acknowledged identification between Victorian literary realism and scientific discourse. George Levine, for instance, asserts that "the epistemology that lay behind realism was empiricist, with its tendency to value immediate experience."[7] He further suggests that realism was "a method consonant with empirical science in that it was exploratory rather than definitive," implying that the realist novel itself could serve as a scientific experiment of sorts.[8] Lawrence Rothfield, meanwhile, has argued that realist authors and nineteenth-century scientists shared a common commitment to mimetic representation. In Rothfield's words, novels like George Eliot's *Middlemarch* (1871–2) approach medical objectivity with their "eschewing of supernatural explanation," "appeal to scientific standards of truth," and "reliance on empirical detail."[9]

The omniscient third-person narration present in many realist novels likewise resembles the objective, scientific detachment of Victorian medical writing. The realist narrator can even be likened to a physician who takes account of characters' "symptomatic" behaviors as part of a detailed character study.[10] Émile Zola, for example, memorably compared the methods of the "experimental" novelist to those of French physiologist Claude Bernard: "This dream of the physiologist and the experimental doctor is also that of the novelist, who employs the experimental method in his study of man as a simple individual and as a social animal."[11] The realist or naturalist author, Zola implied, must "dissect piece by piece" the characters described in his fiction in order to show how heredity and environment determine their actions.[12]

Recent scholarship focusing specifically on the intersections between Victorian physiological psychology and literature has likewise tended to focus on high realist fiction at the expense of genres with less academic prestige, including Gothic novels and romances. For instance, Rick Rylance's impressive *Victorian Psychology and British Culture, 1850–1880* (2000) provides valuable insight into the associationist theories of Alexander Bain, Herbert Spencer, and George Henry Lewes. But when Rylance relates these theories to Victorian literature, he focuses almost exclusively on the work of George Eliot, due to his "respect for her way of dramatizing multiple perspectives."[13]

Similarly, Nicholas Dames's recent study, *The Physiology of the Novel: Reading, Neural Science, and the Form of Victorian Fiction* (2007), performs a valuable service by drawing attention to a previously overlooked strand in Victorian novel criticism. Dames relates how authors like G. H. Lewes, E. S. Dallas, and Alexander Bain examined readers' physiological responses to a novel's pacing, temporal rhythms, and affective qualities. These brain-based reader response theories privileged metaphors of "sound rather than sight" and musical analogies, frequently "borrowing the terms ('movement,' 'theme') of symphonic composition."[14] Dames's work reveals that the recent trend of cognitive literary criticism is not without precedent, and might correctly be viewed as a continuation of an earlier strand of Victorian novel theory. Although the theories Dames describes could reasonably be applied to any novelistic genre, his examples (like Rylance's) are drawn primarily from the canon of high realist fiction, including Eliot's *Daniel Deronda* (1876) and William Makepeace Thackeray's *Vanity Fair* (1847–8).[15]

These recent studies have contributed much to our understanding of Victorian novels by illuminating how medical writing and realist fiction utilize complementary modes of representation. By choosing to focus on realism, however, critics have left the affinities between Victorian physiological psychology and other popular novelistic genres relatively underexplored. In this study, I argue that late-Victorian Gothic novels and romances possessed unique advantages in grappling with certain scientific subjects, specifically, the study of brain structure and function. In so doing, I build upon the work of several recent critics who discuss the role of science in nineteenth-century popular fiction. For instance, Julia Reid has argued that discourses on evolution and degeneration influenced the development of late-Victorian romances, especially those of Robert Louis Stevenson, while Kelly Hurley and Robert Mighall have explored convergences between the Gothic and *fin-de-siècle* discourses on evolution, anthropology, and sexology.[16] Dames and Jenny Bourne Taylor, meanwhile, have discussed readers' physiological responses to mid-Victorian sensation novels, a genre closely related to the Gothic in that both share convoluted plots and a general atmosphere of mystery.[17] However, no one has yet inquired why the late-Victorian Gothic novel and the romance might be ideal mediums for exploring specifically neurological quandaries, particularly the controversies surrounding cerebral localization.

To cite only two examples of this overlap between the neurological and the Gothic, I will turn briefly to the writings of Robert Louis Stevenson and Bram Stoker, the subjects of Chapters 1 and 2, respectively. Both authors received scientific training and later penned famous Gothic romances that hinged upon specific developments in cerebral localization. Stevenson, who briefly studied engineering before deciding upon a literary career, loosely based *Strange Case of Dr. Jekyll and Mr. Hyde* on two famous French case studies of dual personality whose "double lives" were widely discussed in French and British periodicals. In the late nineteenth century, cases of dual personality were often attributed to bilateral brain hemisphere asymmetry. Victorian physiologists like Henry Holland, Arthur Wigan, and Charles Édouard Brown-Séquard argued that if one brain hemisphere were larger than the other, madness and criminality could result. Moral depravity and intellectual regression stemmed from an oversized right brain hemisphere, which supposedly housed primitive instincts and emotions (in stark contrast to the highly evolved left brain). While Jekyll demonstrates rational, civilized, left-brain tendencies, his double, Hyde, exhibits atavistic traits and base passions characteristic of right-brain dominance. In physiological terms, then, Dr. Jekyll is guilty of allowing his right-brain tendencies to overwhelm his more highly evolved left-brain functions.

Bram Stoker, meanwhile, came from a family of successful physicians and obtained a master's degree in science from Trinity College, Dublin.[18] His famous Gothic romance, *Dracula*, contains references to prominent Victorian physiologists such as Ferrier and Burdon-Sanderson, not to mention Jean-Martin Charcot and William Carpenter. Moreover, Stoker's manuscript notes for *Dracula* demonstrate that he sought medical advice on head injuries from his brother, distinguished physician Sir William Thornley Stoker, then president of the Royal College of Surgeons in Ireland. This advice was used to lend clinical accuracy to the death scene of the insect-eating madman, Renfield.

It should come as no surprise, therefore, that *Dracula* tackles some of the thorniest issues raised by cerebral localization debates. Stoker's famous vampire and his minions exhibit semi-conscious, trance-like behaviors that owe much to late-Victorian interest in cerebral automatism and unconscious cerebration. According to localizationists, semi-conscious reflex behaviors such as Lucy's sleepwalking, Dracula's daytime hibernation, and Mina's clairvoyant trances could be traced back to the brain stem. The horror of Stoker's *Dracula* proceeds not just from the Count's repellent vampirism, but also from the looming threat that human beings might be soulless machines governed solely by physiological impulses.

Educated readers of Stoker and Stevenson often saw past the monstrous and supernatural aspects of their tales and grasped the science upon which they were based. These literary works even influenced late-Victorian scientific discourse. For instance, Stevenson's fictional "case" colored scientific work on multiple personality disorder written during the 1880s and 1890s by psychical researcher Frederic Myers and Scottish psychiatrist Lewis Bruce.[19] These examples demonstrate that physicians, scientists, and experts in what we would now call pseudosciences were acutely responsive to literary authors.[20] Rather than a one-way exchange of information between science and literature, Victorian intellectual culture permitted a dialogic or circular conversation in which scientific researchers and literary authors were mutually responsive to one another.

## NEUROLOGY AND THE GOTHIC

Both the Gothic novel and the romance are highly contested genres about which much has been written in the last several decades. As Ian Duncan explains, critics over the last fifty years have applied the label "romance" to many different things, including (but not limited to): "a courtly or chivalric fiction of the late Middle Ages, a fanciful or erotic sentimental enhancement of a situation or event, any unlikely story, highly conventionalized mass-market novels read by women, a narrative with a quest in it, four of the last plays of Shakespeare."[21] The word "Gothic," meanwhile, was used in the eighteenth century to denote Teutonic origins, and also to suggest the medieval or barbaric. Since then, the term has been applied to an architectural style, an aesthetic, a mode, and a poetics, as well as to a literary genre.

In this volume, I will refer to both the romance and the Gothic novel as distinct literary genres, while acknowledging their frequent overlap. When used in a literary context, the label "Gothic" usually applies to a type of popular fiction that flourished between the 1760s and 1820s, beginning with Horace Walpole's *The Castle of Otranto* (1764) and culminating in the suspenseful narratives of Ann Radcliffe. These novels were typically set in gloomy castles or remote convents with labyrinthine passageways, sliding panels, and hidden dungeons; their thematic elements generally included a pervasive atmosphere of darkness and gloom, a complex and attractive villain, supernatural events, and an emphasis on the ancestral past. In terms of form, these novels often featured convoluted or fragmentary narrative structures, which served to augment the reader's pleasurable suspense.[22] While some critics apply the term "Gothic" only to novels written in late-eighteenth- and early-nineteenth-century Britain, others have expanded its application to the horror fiction of other nations and time periods, including "American Gothic" fiction by Edgar Allan Poe and Nathaniel Hawthorne, along with late-Victorian works such as *Dracula* and *Jekyll and Hyde*. These late-Victorian Gothic fictions share many thematic and formal traits with their Romantic-era forebears, but tend to place greater emphasis on monsters (vampires, mummies, etc.) and on contemporary scientific discourses, particularly those relating to evolution and degeneration.

The romance, meanwhile, is a literary genre concerned with the long ago and far away, and with artifice and idealism rather than mimesis. Critics have frequently suggested that the romance has a special relationship with the ancestral past, as befits the genre's distant origins in twelfth-century tales of chivalry and courtly love.[23] In the eighteenth century, the term "romance" was appropriated by writers of Gothic fiction like Walpole and Radcliffe, who wished to signal the distance between their fictions and the commonplace, everyday world of their readers. Yet the romance remained a less prestigious and distinctly feminized genre during the eighteenth century, partly due to its association with Radcliffe. In the early nineteenth century, Walter Scott helped to remasculinize and legitimize the romance with tales of adventure such as *Waverley* (1814) and *Ivanhoe* (1819).[24] By the time of the romance revival of the 1880s and 1890s, writers like Stevenson, H. Rider Haggard, and Andrew Lang conceived of the romance as a distinctly masculine form whose artistry rivaled that of the realist novel.[25]

The above histories suggest how, well before the late-Victorian period, the boundaries of the Gothic and the romance had already begun to blur. Even high realist novels have moments of Gothic mystery: take, for instance, the "dream-like strangeness" of Dorothea's honeymoon in *Middlemarch*, or the catastrophic, highly implausible flood that concludes Eliot's *The Mill on the Floss* (1860).[26] Moreover, hybrid fictions like those of Walpole and Radcliffe introduced new issues of terminology. Some critics call such works "Gothic romances," while others suggest that this phrase is redundant, since the Gothic is arguably a subset of the romance.[27] Still others use the terms "Gothic" and "romance" almost interchangeably, ignoring the important differences between the two genres (such as the idealistic, utopian tendencies of the romance). In this study, I often apply the label "Gothic romance" to works that seem to fit both generic categories. For instance, *Jekyll and Hyde* or *The Island of Doctor Moreau* (1896) might be referred to as Gothic romances, but not Corelli's *A Romance of Two Worlds* (1886), which is notably free from Gothic elements.

These generic boundary lines become even more difficult to assess in late-Victorian and Edwardian fiction, despite – or perhaps because of – the resurgent popularity of Gothic novels and romances during this time. Sensational tales of adventure and horror proliferated between 1880 and 1914, due to shifting artistic tastes and sweeping social changes. Scholars of Victorian fiction continue to debate the usefulness of the label "Gothic" as applied to this body of literature. Nicholas Daly points out, for instance, that Victorians themselves did not use this term. Instead, they referred to works like *Dracula* and *Jekyll and Hyde* simply as "romances." Nevertheless, Daly concedes that a novel like *Dracula* shares numerous formal and thematic elements with Radcliffe's *The Mysteries of Udolpho* (1794).[28] At the other extreme, several critics have vastly expanded the range of late-Victorian fiction to which we might apply the "Gothic" label.[29]

In *Popular Fiction and Brain Science in the Late Nineteenth Century*, I will explore how certain formal and thematic conventions of the Gothic and romance genres mesh surprisingly well with a certain non-evolutionary strand of Victorian scientific thinking. When previous scholars have explored resonances between late-Victorian science, the romance, and the Gothic, they have tended to focus on those branches of science most closely associated with evolution, such as degeneration theory, criminology, anthropology, sexology, evolutionary psychology, and so forth. Cases in point include Hurley's *The Gothic Body: Sexuality, Materialism, and Degeneration at the* Fin de Siècle (1996), Mighall's *A Geography of*

*Victorian Gothic Fiction: Mapping History's Nightmares* (1999), and Julia Reid's *Robert Louis Stevenson, Science, and the* Fin de Siècle (2006).

By focusing specifically on neurology in relation to Gothic novels and romances, *Popular Fiction and Brain Science in the Late Nineteenth Century* positions itself somewhere between the social science or "soft science" milieu of these earlier studies and the more prestigious "hard sciences," like physics and astronomy, whose practitioners communicated primarily via mathematics rather than vernacular languages. According to Alice Jenkins, the nineteenth century was the period in which "hard" versus "soft" sciences came to be defined against one another, with hard sciences like physics offering a promise of mathematical rigor to which other sciences aspired.[30] Neurology took part in this larger development in scientific discourse, even if it did not (and still does not) fall neatly on either side of the hard versus soft divide. During the second half of the nineteenth century, neurologists increasingly incorporated mathematics, charts, graphs, and specialized jargon in their articles and books, making them less accessible to laypeople. This was especially true after 1891, when Santiago Ramón y Cajal's neuron doctrine took hold in Europe. Thereafter, localizationists increasingly debated what occurred in the brain at the cellular level, investigating matters of anatomy, cell biology, and microscopic electrical and chemical reactions.

At the same time, Victorian neurologists were well aware that their work reflected on broader issues that were not so easy to quantify, particularly those relating to biological determinism. Implicitly or explicitly, neurological research posed the following questions: what is the role of the will or soul in human action? Where does strictly physiological activity leave off, and a higher power take over? When does a human being resemble a machine more than a self-determining organism? These questions put pressure upon the boundaries between human and divine, human and animal, human and machine. Boundaries (and their transgression) likewise hold a central place in Gothic fiction from Walpole onward, as numerous critics have elucidated. Eugenia DeLamotte argues that eighteenth-century Gothic romances generated terror by violating "the boundaries of the self," often by means of "transgression[s] against the body, the last barrier protecting self from other."[31] Similarly, Stephen Bruhm demonstrates how Romantic-era Gothic fiction emphasized the limits of the body's endurance, as well as the boundary between self and other that is invoked when we witness or read about another person's pain.[32] The late-Victorian Gothic romances examined here test a different, very specific set of limits, the most important of which may be the boundary between voluntary and reflex action. When this boundary becomes unclear – as when Jekyll changes spontaneously into Hyde, or when Dracula's victims feel instinctive attraction to the loathsome vampire – the novel's characters experience affective responses ranging from fear and anger to a sense of betrayal or abandonment by God.

Many Victorians experienced these same emotions upon first encountering cerebral localization research. In fact, if a scientific discourse can be said to have a mood or tone, late-Victorian neurology could justly be characterized as a Gothic science. This statement might seem to jar with the widespread perception that science is unemotional or impartial. Mighall observes, for instance, that "horror fiction has a generic obligation to evoke fear or suggest mystery," whereas "science … attempts to contain fear and offer a rational explanation for all phenomena."[33] This remark seems applicable to some Gothic romances, like *Dracula*, where an authorial pose of scientific objectivity allows the reader to approach horrifying supernatural subjects with relative calm.

But Mighall's commentary overlooks the disturbing, Gothic undertone of some late-Victorian science. For instance, *fin-de-siècle* degeneration theories, which derived indirectly from Darwinian evolutionary thought, heralded an imminent biological apocalypse that would culminate in the extinction of mankind.[34] Victorian sexologists, meanwhile, defined normal sexual functioning against the backdrop of numerous perversions, including Richard von Krafft-Ebing's concept of the *Lustmord* or "Lust Murder" (in which killing replaces the sexual act).[35] As these examples suggest, Gothic novels drew some of their horrifying elements directly from contemporary science. Such fictions served both to aggravate and alleviate anxieties generated by nineteenth-century biology, evolutionary theory, sexology, and criminal anthropology, as Hurley has explained.[36] The combined impact of these scientific discourses destabilized prevailing ideas about what it meant to be human.

One might easily add neurology – specifically, cerebral localization theories – to the list of *fin-de-siècle* scientific discourses that undermined a sense of a stable human identity. By suggesting that certain parts of the brain controlled specific emotions and behaviors, localizationists contradicted the popular belief in a unified soul or mind governing human action, thus narrowing possibilities for human agency. Even within the scientific community, controversy brewed regarding this seemingly mechanistic view of the mind. In 1846, French physiologist Jean Pierre Marie Flourens argued that pinpointing the cerebral origin of movements and thoughts apparently "undermin[ed] the unity of the soul, human immortality, free will, and the very existence of God."[37] He was far from alone in this view. Three decades later, eminent physiologist William Benjamin Carpenter surveyed the field of psychology, and lamented that "the inevitable conclusions of Physiological science are now advanced in proof of the Determinist hypothesis."[38] He was frankly distressed by the idea that seemingly "purposive" mental actions might be attributed to "nothing else than a physical mechanism."[39]

The pseudoscientific forerunners of cerebral localization, phrenology and phreno-mesmerism, proved equally disturbing to late-eighteenth- and early-nineteenth-century views of man's place in the universe.[40] Phrenology, an early and rather clumsy attempt to correlate emotions, abilities, and behaviors with specific brain regions, was invented in the late eighteenth century by Austrian physician Franz Joseph Gall (1758–1828). Gall convinced the scientific community that the brain was the organ of mind, a previously controversial notion. He further contended that the size of different regions of the brain was a reliable indicator of an individual's talents and personality traits.[41] Since the skull takes its shape from the brain, Gall surmised, the surface of the skull could be read as an accurate indicator of aptitudes and tendencies. Gall's ideas rapidly caught on in Europe and spread to England and America via phrenological disciples like Johann Gaspar Spurzheim, George Combe, and O. S. Fowler. Phrenology and related discourses like phreno-mesmerism remained popular in England until mid century, while in America they held sway even longer.[42]

Despite the widespread popular and scientific interest generated by Gall's theories, their originator was treated as a heretic. Gall was refused a Catholic burial and his works were placed on the Catholic *Index* of forbidden works because his ideas (particularly his assertion that the brain was the organ of mind) felt disturbingly materialist to his contemporaries.[43] Severe as this reaction may seem, it was a predictable response to ideologically threatening concepts. As Alan Richardson has explained, phrenology and other Romantic-era theories of brain structure and function incited controversy because they called into question "the existence of the soul, the necessity of God, and the integrity of the self."[44]

Like late-Victorian cerebral localization theories, phrenology was often explored and exploited in Gothic fiction, beginning with Charles Brockden Brown in the late eighteenth century and continuing well into the mid-Victorian era.[45] Edgar Allan Poe critiqued phrenological theories in short pieces such as "The Imp of the Perverse" (1845), while employing phrenological description in detective stories and Gothic tales such as "The Murders of Rue Morgue" (1841) and "The Fall of the House of Usher" (1839).[46] Phrenological readings also figure prominently in mid-Victorian Gothic narratives like Charlotte Brontë's *Jane Eyre* (1847) and *Villette* (1853), not to mention George Eliot's sole foray into Gothic fiction, *The Lifted Veil* (1859).[47]

Whereas phrenology was a dubious pseudoscience whose practitioners were frequently accused of quackery, the cerebral localization experiments performed after 1860 by Jackson, Fritsch, Hitzig, and Ferrier were firmly grounded in the scientific method and widely regarded as cutting-edge research. Nonetheless, cerebral localization trailed an odor of Gothic mystery left over from its pseudoscientific predecessor. Perhaps this is because late-Victorian cerebral localization theories, like phrenological discourse, challenged revered assumptions about the soul, the will, and the nature of God. But the association between cerebral localization and the Gothic may also stem from the abrupt, brutal manner in which many laypeople first confronted cerebral localization theories and experiments in 1881.

In that year, famed neurologist David Ferrier was put on trial for allegedly violating the 1876 Anti-Vivisection Act. Ferrier's trial was extremely well publicized, drawing unprecedented public attention and scrutiny to neurological theories and experimental methods. Since 1873, Ferrier had been performing experiments in which he applied electrical currents to the brains of live monkeys, cats, and dogs. He then studied changes in the animals' behavior once they awoke from anesthesia. As related above, the cortical maps Ferrier developed as a result of these experiments proved to be lifesaving medical breakthroughs.[48] But to many members of the public, and to antivivisection activists in particular, Ferrier's experiments seemed unimaginably cruel. One opponent was Frances Power Cobbe, the leader of the antivivisectionist Victoria Street Society (the entity responsible for hauling Ferrier into court). In *The Modern Rack: Papers on Vivisection* (1889), Cobbe described the neurologist's experiments in frighteningly graphic terms: "The experiments of Ferrier on monkeys and of [German physiologist Friedrich] Goltz on the brains of dogs involve different mutilations, with scooping out of the brains, till, in some cases, they resemble, as Goltz has said, a 'lately-hoed potato-field.'"[49]

Antivivisectionists felt repulsed not only by the visceral details of the experiments, but also by their philosophical ramifications. That human brain function could be predicted on the basis of animal experimentation irrefutably demonstrated the similarity between men and beasts, reaffirming the disturbing conclusions of Charles Darwin's *The Origin of*

*Species* (1859) and *The Descent of Man* (1871). More troubling still, Ferrier's ability to produce complex behaviors by applying electrical current to the brain suggested, in Laura Otis's words, that "there was nothing sacred about the human will, not even human consciousness."[50] After the trial, Gothic novels like Wilkie Collins's *Heart and Science* (1883) and H. G. Wells's *The Island of Doctor Moreau* invoked the specter of the Ferrier trials to generate fascination and horror.[51] Although Ferrier was eventually acquitted, the legacy of his trial was the association of cerebral localization with inhumane experimental methods and disturbing philosophical conclusions.

If the Gothic helped authors and readers come to terms with disturbing implications of scientific discourses, as Hurley argues, then it makes sense that writers like Stoker, Stevenson, and Wells turned to this genre while grappling with the fallout of neurological experiments. But there are other plausible reasons to link the Gothic with uncharted territories of the human brain. Perhaps surprisingly, one such argument came from the pen of psychologist William James, whose brother, arch-realist Henry James, famously contended with Stevenson over the relative merits of romance versus realism.[52] William James felt that nature was too baroque to be fairly depicted by linear narratives: "Nature is everywhere gothic, not classic. She forms a real jungle, where all things are provisional, half-fitted to each other, and untidy."[53] Scientific writing, he suggested, could force nature into uncomfortable linear narratives that seemed, at best, an imperfect fit. This tendency could be especially problematic when science treated tangled subjects like human psychology and brain function.

James's critique of scientific rationalism is most pointedly expressed in his writings on psychical research, particularly his 1901 essay, "Frederic Myers's Service to Psychology." In this piece, James usefully distinguished between "classic-academic" and "romantic" types of imagination, arguing that the linear scientific reasoning of traditional psychology should be combined with more intuitive, imaginative approaches to the human mind:

> The [classic-academic imagination] has a fondness for pure clean lines and noble simplicity in its constructions. It explains things by as few principles as possible and is intolerant of either nondescript facts or clumsy formulas. The facts must lie in a neat assemblage, and the psychologist must be enabled to cover them and "tuck them in" as safely under his system as a mother tucks her babe in under the down coverlet on a winter night. Until quite recently all psychology, whether animistic or associationistic, was written on classic-academic lines. The

> consequence was that the human mind as it figured in this literature, was largely an abstraction. A sort of sunlit terrace was exhibited on which it took its exercise. But where the terrace stopped, the mind stopped …
>
> But of late years the terrace has been overrun by romantic improvers, and to pass to their work is like going from classic to Gothic architecture, where few outlines are pure and where uncouth forms lurk in the shadows. A mass of mental phenomena are now seen in the shrubbery beyond the parapet. Fantastic, ignoble, hardly human, or frankly non-human are some of these new candidates for psychological description. The menagerie and the madhouse, the nursery, the prison, and the hospital, have been made to deliver up their material. The world of mind is shown as something infinitely more complex than was suspected; and whatever beauties it may still possess, it has lost at any rate the beauty of academic neatness.[54]

Here, James was specifically discussing the work of Myers, whose "romantic improvements" to the science of psychology included systematic inquiries into the possible existence of ghosts, spirit rapping, telepathy, and other forms of extrasensory perception. But James's Gothic or romantic imagination could just as easily be ascribed to writers of fiction like Stevenson, Stoker, or Wells, who explored human psychology in all its troubling complication, without the artificially imposed linearity of mainstream scientific discourses. These authors' monstrous creations numbered among the "Fantastic, ignoble, hardly human, or frankly non-human" subjects that mainstream science approached obliquely, if at all. Writers of Gothic fiction, like psychical researchers, unearthed the dark, tangled corners of the mind that seemingly had no place on Ferrier's maps of the cerebral cortex.

Nearly a century after James's critique of "linear" science, neuroscientists Oliver Sacks and A. R. Luria made similar observations about the tendency toward simplification and schematization that dominates – and arguably impoverishes – present-day neurology. Ever since Jackson and other localizationists inaugurated the "classical" era of neurology in the 1860s, neurologists have focused primarily on left-hemisphere brain function, overlooking the more baroque and puzzling disorders afflicting the right-brain hemisphere.[55] Sacks explains that "the left hemisphere, like a computer … is designed for programs and schematics, and classical neurology was more concerned with schematics than with reality." By contrast, the right-brain hemisphere "controls the crucial powers of recognizing reality which every living creature must have in order to survive," including facial recognition, proprioception, non-verbal communication, and other functions not easily approximated by mathematical or computational models.[56]

In taking a computational rather than holistic view of brain function, Sacks argues, neurologists effectively ignore the entire right hemisphere, and with it "the physical foundations of the *persona*, the self."[57] In order to correct this fundamental imbalance, Luria called for (and occasionally practiced) a "romantic" science based in narrative case histories that could explore right-hemisphere disturbances in all of their complexity. In books like *The Mind of a Mnemonist* (1968), Luria explored how one aberrant psychological trait (say, an extraordinarily accurate visual memory) affects an individual's "total personality structure," including his functioning in career and family life.[58] Sacks would later pursue a similar line of inquiry in books like *The Man Who Mistook His Wife for a Hat and Other Clinical Tales* (1985), which contains narrative case histories of several patients with right-hemisphere disorders.

To be sure, some nineteenth-century case studies more closely resemble Luria or Sacks's ideal of romantic science than today's jargon-laden medical reports. Take, for instance, the confessional case reports of sexologists such as Richard von Krafft-Ebing and Havelock Ellis, or Sigmund Freud and Josef Breuer's exploration of the traumatic origins of hysteria.[59] But in general, nineteenth-century case studies shifted toward a clinical, impersonal format as scientific disciplines became increasingly professionalized. Historians of medicine have suggested that from the mid eighteenth century onwards, medical professionals increasingly privileged visual evidence and measurable data over personal anecdotes and patient testimony.[60] Lilian Furst explains how these tendencies crystallized in twentieth-century medical reports, some of which are "written as though the patient did not exist."[61]

This shift toward objectivity and linearity was perhaps more apparent in late-Victorian neurology than in some other medical fields. Localizationists relied primarily on data obtained from postmortem examinations and animal experiments, making patient experience less relevant. This bias is apparent in neurological case studies of the era, particularly those published in avant-garde journals such as *Mind: A Quarterly Review of Psychology and Philosophy* (1876–present) or *Brain: A Journal of Neurology* (1878–present). For instance, Ferrier's case report, "The Brain of the Criminal Lunatic" (1882), relates potentially sensational subject matter in a detached, scientific manner. In this article, Ferrier describes how he "examined the brain of a woman who in 1871 murdered two of her children, but who … was found insane and incapable of the charge against her."[62] The woman's extraordinary case history takes up only the first three pages of the article, while the remaining ten pages focus on the condition of her brain as revealed during a postmortem examination.

Ferrier relates his findings in a detached, technical manner obviously geared toward his scientific peers rather than a general audience. Upon examination, he found the woman's right brain hemisphere to be substantially larger and more developed than the left. According to prevailing medical wisdom at the time, such a hemispheric imbalance could cause criminality and insanity. Ferrier's concluding remarks emphasize the degree to which his clinical findings cohere with prevailing scientific ideas: "considered in reference to the symptoms, the condition of the brain is entirely in harmony with them, according to the doctrines of localization maintained by myself and others."[63] Notably, the patient herself has been reduced to a "brain" whose only significance is its contribution to localization research.

While a scientist might laud Ferrier for his rigorous objectivity and professionalism, a novelist would sense a missed opportunity to explore this woman's unusual mental state, not to mention her criminal activities. Four years after Ferrier's study appeared in *Brain*, Stevenson's *Jekyll and Hyde* fleshed out the narrative possibilities of this case (or one very like it) by detailing the experiences of a criminally insane doctor suffering from hemispheric imbalance. The novella concludes with "Henry Jekyll's Full Statement of the Case," which provides the autobiographical "testimony" of this imaginary sufferer of neurological disease. *Jekyll and Hyde* thus demonstrates how Gothic fiction simultaneously complements and critiques classical, post-Jacksonian neurology by focusing on subjective experience rather than objective data. Like the romantic science of Sacks or Luria, then, Victorian Gothic fiction could add psychological depth to dry scientific ideas, and suggest greater narrative possibilities for the medical case study.

Victorian Gothic romances contrast with post-Jacksonian neurology in form as well as content. In contrast to the oversimplifying linearity of classical neuroscience, Gothic prose is snarled by multiple narrators, embedded texts, instances of doubling and mistaken identity, and numerous indications of narrative instability and unreliability. The realist novel, like a scientific narrative, typically adheres to a linear plot structure and employs a distancing, purportedly objective omniscient narrator. By contrast, the romance or Gothic novel opens up more immediate, sensational, and baroque possibilities. In other words, it is no coincidence that the complicated narratives, subterranean passages, and involved storylines traditionally associated with the late-Victorian Gothic subtly remind us of the convoluted surfaces of the brain.

## NEUROLOGY AND THE ROMANCE

While the Gothic thus presented certain advantages in grappling with *fin-de-siècle* neurology, so did the romance, the favorite genre of Stevenson, Corelli, and Wells, among others. The work of these three authors exemplifies how the Gothic novel and the romance overlapped with the emergent genre of science fiction. Although the term "science fiction" was unavailable to late-Victorian authors, the tradition that now goes by that name arguably began with the Faust legends of the sixteenth century and continued with Mary Shelley's seminal Gothic novel, *Frankenstein* (1818).[64] Wells's early scientific romances and Stevenson's *Jekyll and Hyde* fit the Frankensteinian mold by offering a combination of romantic, Gothic, and science fiction tropes. Corelli's works likewise unite elements of the romance and science fiction, while simultaneously responding to the anxieties that pervade *fin-de-siècle* Gothic novels such as *Dracula* and *Jekyll and Hyde*.

These generic overlaps are no coincidence. Brian Aldiss writes that "science fiction was born from the Gothic mode, and is hardly free of it now."[65] Patrick Brantlinger suggests, meanwhile, that both the Gothic and science fiction are "forms of apocalyptic nightmare fantasy" that warn readers about the dangers of science and of "reason taken to extremes."[66] In other words, Brantlinger paradoxically reads science fiction and its predecessors as deeply anti-scientific genres. This position seems entirely tenable for disturbing works like *Jekyll and Hyde* and Wells's *The Island of Doctor Moreau*, with their obvious critiques of scientific hubris, but perhaps less so for light-hearted scientific romances such as Wells's *The Food of the Gods* (1904) or utopian, spiritualist science fiction like Corelli's *A Romance of Two Worlds*. In any case, while I read the fictions of Wells, Stevenson, and Corelli primarily as examples of the romance, their novels amply demonstrate how the works in this study bleed over into multiple generic categories.

The late-Victorian romance revival arguably began in the mid 1880s, as works like Stevenson's *Treasure Island* (1883) and Haggard's *King Solomon's Mines* (1885) signaled a "new direction" in popular fiction.[67] These works, along with the James–Stevenson debates on the "Art of Fiction," helped solidify the public perception that the romance, with its associations of wish-fulfillment, escapism, and fantasy, stood opposed to the realist novel, with its emphasis on truthful representation.[68] The late-Victorian romance exploited exotic settings, fast-paced, plot-driven narratives, and motifs of imprisonment, mazes, or entrapment. Additionally, romances often invoked some sort of "break in ordinary consciousness" signaled by a shipwreck or quasi-magical transformation.[69]

With several prominent exceptions, recent critics have tended to overlook the late-Victorian romance as a serious art form.[70] This oversight is due in part to the legacy of modernist literary critics, who dismissed the romance as a highly commodified form.[71] There is also a widespread awareness that this genre, particularly in the hands of lesser practitioners like Haggard or Anthony Hope, could serve as a vehicle for misogynist or imperialist sentiment.[72] For a skilled and scientifically educated author like Stevenson, however, the romance bodied forth elements of late-Victorian evolutionary psychology, as Reid has shown. In contrast to W. D. Howells and Henry James, who lauded realist novels for their intellectual sophistication, Stevenson celebrated the romance for its appeal to ancestral memory and its evocation of atavistic instincts and pleasures that are too often suppressed by modern living. The romance, Stevenson argued, spoke to the savage within us: a primitive, ape-like ancestor who was "probably arboreal" in his habits.[73] While critics then and now have frequently dismissed Victorian romances as one-dimensional adventure stories for boys, Reid demonstrates how Stevenson's romances expressed the author's nuanced engagement with late-Victorian evolutionary psychology.

The romance resonated not just with Victorian evolutionary ideas, but also with neurological concepts. Stevenson's arguments in favor of the romance hinged partly on its nervous impact on the reader, whose "civilized nerves still tingle … with rude terrors and pleasures."[74] Like the Gothic novel or the sensation novels popular in the 1860s, the romance appealed to readers and aroused controversy due to its effects on the nervous system.[75] The sensation novel, for instance, was frequently criticized for arousing distressing (or inappropriately pleasurable) somatic responses in the reader, such as "Making the Flesh Creep, Causing the Hair to Stand on End, [and] Giving Shocks to the Nervous System."[76]

The romance, Stevenson contended, likewise involved the reader in "an unbroken round of pleasure and suspense" that could, and should, have an immediate and lasting nervous impact.[77] In "A Humble Remonstrance" (1884), Stevenson's rejoinder to Henry James's "The Art of Fiction" (1884), Stevenson described the emotional and physiological effects of the "novel of adventure": "Danger is the matter with which this class of novel deals; fear, the passion with which it idly trifles."[78] The novel is successful, he argued, "if our pulse be quickened."[79] In "A Gossip on Romance" (1882), meanwhile, Stevenson described how a successful romance should arrest and hold our attention by arousing our senses:

> We should gloat over a book, be rapt clean out of ourselves, and rise from the perusal, our mind filled with the busiest, kaleidoscopic dance of images, incapable of sleep or of continuous thought. The words, if the book be eloquent, should run thenceforward in our ears like the noise of breakers, and the story, if it be a story, repeat itself in a thousand coloured pictures to the eye.[80]

In other words, late-Victorian romances, with subject matter ranging from adventure on the high seas to spine-tingling monstrosities, aimed to provoke an immediate, visceral reader response – specifically, a *nervous* response appropriate to the neurological subject matter these romances often addressed.

The affinity between neurology and the romance emerges clearly in *Jekyll and Hyde*, Stevenson's most profound engagement with late-Victorian cerebral localization. Not only does the 1886 "shilling-shocker" pleasurably jangle readers' nerves, it also immerses us in a detailed exploration of bilateral brain hemisphere asymmetry and its possible effects. Moreover, this particular Gothic romance became a means for Stevenson to simultaneously mimic and critique the apparent objectivity of the medical case study, as I will discuss at length in Chapter 1.

H. G. Wells's immense popularity as an author of science fiction has often obscured the other accomplishments of his varied career, including his innovations in the form of the late-Victorian romance. Wells viewed the romance as a "mouthpiece for science," as he stated in 1895, and as an imaginative means of testing scientific hypotheses.[81] In the introduction to a 1933 edition of his so-called "scientific romances," Wells described his method of focusing on scientific "impossibilities" played out in realistic, everyday settings.[82] In works like *The First Men in the Moon* (1901) and *The Time Machine* (1895), Wells concentrated on one hypothetical proposition, such as "how would you feel and what might not happen to you … if for instance pigs could fly and one came rocketing over a hedge at you? … or if you became invisible?"[83] Wells then isolated this imaginative "what if" question and made all other aspects of the story as realistic and commonplace as possible. He aimed, in other words, to "*domesticate* the impossible hypothesis" so as to trick the reader into suspending disbelief.[84]

Wells's distinctive manner of crafting a romance strikingly resembles the scientific method, wherein one variable is tested against a series of controls. Perhaps this likeness should not surprise us, coming from an author who studied under Darwin's apologist, Thomas Huxley, at the Normal School of Science in South Kensington. Although Wells's experimental novelistic technique might initially resemble that of a realist or naturalist like Zola, the difference lies in the fanciful nature of the variable tested, such as the possibility of time travel or human invisibility. Moreover, the results of Wells's "experiments" are often strikingly Gothic, such as the near extinction of mankind depicted in *The War of the Worlds*.

In early scientific romances like *The War of the Worlds* and *The First Men in the Moon*, Wells tested a specific neurological hypothesis with results both comic and terrifying. Wells was fascinated by the Lamarckian idea of unchecked brain evolution occurring at the expense of bodily strength and stamina. According to Lamarck's first law, expounded in *Zoological Philosophy* (1809), organs that are frequently used tend to develop more quickly and hence to grow larger, while little-used organs wither away and ultimately disappear. This logic still resonated with many *fin-de-siècle* scientists, for whom Darwin's theory of natural selection had not entirely supplanted the Lamarckian concept of inheritance of acquired characteristics.[85] Wells imaginatively brought this Lamarckian evolutionary hypothesis to life in top-heavy creations like The Grand Lunar of *The First Men in the Moon*, whose enormous cerebrum dwarfs his stunted body, and the large-brained, merciless aliens of *The War of the Worlds*. These extraterrestrials' brains have rapidly evolved at the expense of their dwindling bodies – an evolutionary future Wells likewise envisioned for mankind, as he hinted ominously in "The Man of the Year Million" (1893).

While Wells and Stevenson wrote shudder-inducing dystopian tales that might reasonably be called Gothic romances, Marie Corelli's *A Romance of Two Worlds* presents a utopian blending of religious and neurological ideas. In this now forgotten Victorian blockbuster, Corelli's nameless heroine overcomes neurasthenia or "nerve weakness" by harnessing electrical forces housed in the spiritual entities that surround her. Using her new reserves of nervous force, Corelli's heroine travels through outer space to view the planets and even heaven itself. In later romances such as *The Life Everlasting* (1911), Corelli incorporates Cajal's neuron doctrine into her eclectic spiritual philosophy, suggesting that neurons are tiny storehouses that preserve God's heat, light, and love. By fully accessing the spiritual energies present in our neurons, Corelli implies, humans might be able to heal the sick, remain eternally youthful, and communicate directly with God.

Whereas Wells and Stevenson used the romance to pleasurably jar our nerves, the calming rhythms of Corelli's fiction have the opposite effect, soothing the reader's fears and simulating the joys of spiritual harmony. In an 1886 letter to publisher George Bentley, Corelli wrote, "as far as the public goes, I want to make it also feel rested, invigorated, and rendered for a time oblivious of its troubles in the perusal of my paper."[86] This escapism was part of Corelli's broader mission, in which the romance could serve as therapy and moral uplift for troubled readers. Corelli hoped that *A Romance of Two Worlds* would soothe "the minds of those unhappy ones who have been unsteadied by modern atheistical books and doctrines, and feel themselves utterly wretched, not knowing where to turn."[87] For Corelli's readers, as for her space- and time-traveling heroines, escape could be a means to religious enlightenment.

Though these three authors employed the romance form quite differently, they all found it congenial to exploring neurological ideas, especially those related to cerebral localization. All of them situated themselves in opposition to realism as practiced and theorized by Henry James, who argued that the aim of the novelist was to "compete with life" by offering a complex, painterly rendition of the social universe he observed around him.[88] Paradoxically, Wells, Corelli, and Stevenson found that departing from such truthful representation into the realm of fantasy allowed them greater freedom to ferret out scientific "truths" about the structure and function of the human brain. For these writers, romances that tested the limits of perception and imagination proved best suited to exploring the unmapped territories of the cerebral cortex.

In short, the engagement with cerebral localization theories in late-Victorian Gothic novels and romances should not strike us as surprising or out of place. While the labyrinthine contours of the Gothic novel proved an ideal medium for exploring the brain's convolutions, the spine-tingling terrors and pleasures of the romance prepared readers to encounter the neurological subject matter these tales frequently broached. The historical associations between phrenology and the Gothic, the journalistic frenzy surrounding Ferrier's 1881 trial, and the frighteningly atheistic philosophical ramifications of neurological discoveries likewise paved the way for a natural association between the late-Victorian Gothic and cerebral localization. Most importantly, the convoluted narratives of the Gothic and the fantastical elements of the romance served as a corrective to the linear worldview and inflexible objectivity of science. Whereas scientists rigidly excluded subjective elements and inconclusive data from their studies, authors of Gothic romances freely explored the dark territories and uncharted regions of the brain.

## BEYOND FEAR?

So far, I have primarily discussed those Victorian authors (Stoker, Stevenson, and Wells) who voiced anxiety about cerebral localization, particularly its moral and theological implications. But the fictions of Corelli and Grant Allen demonstrate that fear was hardly the only viable response to these neurological discoveries. In fact, literary authors' responses to localization ran the gamut from reactionary, to celebratory, to visionary or prophetic.

But the novelists discussed here also felt ambivalence toward the neurological developments dramatized in their fictions, whether they admitted it or not. Those authors who apparently embraced aspects of cerebral localization, such as Allen or Corelli, sometimes unwittingly betrayed signs of unease about the materialistic assumptions of late-Victorian neurology, or attempted to "revise" the science itself to better fit their own worldview. Meanwhile, those authors who were most highly critical of science, such as Stoker and Stevenson, nonetheless appropriated it in ways that acknowledged their complicity with its power and influence – a dynamic akin to what Linda Hutcheon has described as "complicitous critique" within the context of postmodern literature and art.[89] The five chapters of this book are divided up so as to emphasize the differences in opinion between the literary authors discussed here, while underscoring that none of them – not even those devoted to the promotion and professionalization of science – seemed entirely reconciled to these emergent neuroscientific developments.

Paradoxically, some of the most scientifically literate authors discussed in this volume proved the most hostile to cerebral localization. Part I of this book, "Reactionaries," describes the fearful reactions of Stoker and Stevenson toward this relatively new scientific concept. Stoker condemned the biological determinism of localization through the eponymous villain of *Dracula*, a soulless vampire who embodies the perils of automatic brain function unchecked by willpower. Stevenson likewise adopted a cautionary tone throughout *Jekyll and Hyde*, in which the divided protagonist suffers from brain hemisphere asymmetry that requires a chemical cure. In their Gothic romances, Stoker and Stevenson evoked the distressing idea that humans are motivated by reflex actions of the cerebrum that remain outside of conscious awareness, a possibility that still feels provocative and disturbing.

Stevenson and Stoker's fearful reactions to cerebral localization best typify the Victorian public response to localization experiments, which were viewed with suspicion after Ferrier's infamous trial. But some authors reacted quite differently to these neurological discoveries, as the next two parts of this volume emphasize. Part II of this volume, "Materialists," explores how Allen embraced cerebral localization theories by deliberately portraying the heroine of his 1891 novella, *Recalled to Life*, as a human machine. Allen, an Anglo-Canadian popular science writer and novelist, was also an atheistic contrarian who loved thumbing his nose at bourgeois morality. Accordingly, he celebrated mechanical metaphors of brain function in fictions such as *Recalled to Life*. In this underappreciated Gothic detective novella, protagonist Una Callingham is explicitly compared to a camera whose optic nerve takes pictures of her surroundings. Allen further suggests that Una's memory resembles a sequential series of still photographs, not unlike the chronophotography practiced by Eadweard Muybridge and others from the 1870s onward. The use of mechanical metaphors for brain functions like sensory perception and visual memory was a self-consciously radical maneuver, I argue, at a time when comparisons between men and machines provoked fear and protest from religiously conservative Victorians. But at the same time, Allen's heroine possesses a psychological depth that seemingly undermines her status as machine, suggesting that Allen's embrace of materialist psychology was neither as complete nor as successful as he himself believed.

Part III, "Visionaries," contrasts the apocalyptic and utopian romances about the brain written by Wells and Corelli, respectively. These two very different authors took cerebral localization theories in visionary and sometimes frankly occult directions. By juxtaposing Wells and Corelli, I demonstrate how late-Victorian proto-science fiction (like its modern counterpart) swayed between the poles of utopian and dystopian thinking. In scientific romances such as *The War of the Worlds* and *The First Men in the Moon*, Wells depicted giant brains from outer space whose puny bodies and overgrown cerebrums foreshadow possible long-term results of Lamarckian brain development.

Corelli, the most financially successful author of the 1890s, is also the only woman writer in this study and the only novelist who lacked formal scientific training. She viewed neural networks as pathways to God and neurons themselves as storehouses of heat, light, and God's love. Her inspired fusion of cerebral localization with spiritualism and heterodox

Christianity helped shrink the widening gap between science and faith during the late nineteenth and early twentieth centuries. While Stevenson and Wells penned romances foreshadowing the potentially disastrous consequences of neurological discoveries, Corelli's *A Romance of Two Worlds* embraced the radical idea of communing with God through neural channels of communication. Of the writers discussed here, Corelli was the only one who saw spiritual fulfillment as compatible with Victorian neuroscience. Her unprecedented popularity suggests that many late-Victorian readers likewise entertained this view. Yet Corelli interpreted neuron doctrine loosely and creatively; in effect, she had to *revise* the scientific ideas she appropriated in order to successfully incorporate them into her spiritually uplifting fiction. She deliberately denied or ignored the materialist implications of the scientific concepts she co-opted, often leading to misunderstanding or misuse of ideas about radioactivity and communication between neurons.

Despite their differences, the five literary figures featured in this study all found the Gothic novel and the romance congenial to their neurological subject matter. Perhaps this is because late-Victorian neurology itself possessed an aura of Gothic mystery, even an element of horror. During and after the Ferrier trials, the Victorian lay public was rudely confronted with the idea that their mental and spiritual activities might be purely mechanical, traceable to electrical activity in specific regions of the brain. In the wake of this unsettling discovery, readers craved fiction that challenged the content and form of classical neurology. The convoluted narratives and occult subject matter of Gothic romances fulfilled this need, serving as correctives to the terse, linear narration of medical case studies. The novels discussed in this volume all suggest that we are more than the sum of our neuronal activity, more than the interaction of our cerebral hemispheres. And that was exactly the message that Victorian audiences wanted to hear.

# PART I

# *Reactionaries*

CHAPTER I

# *Robert Louis Stevenson's* Jekyll and Hyde *and the double brain*

In "The Decay of Lying: An Observation" (1889, revised 1891), Oscar Wilde wrote that "the transformation of Dr. Jekyll reads dangerously like an experiment out of the *Lancet*."[1] This statement rings true on more levels than Wilde himself probably realized. Not only does Robert Louis Stevenson's *Strange Case of Dr. Jekyll and Mr. Hyde* (1886) resemble Victorian case studies in its form and structure, but its core idea may also have originated from medical literature. In 1905, Fanny Stevenson traced her spouse's interest in dual personality to a specific scientific article: "[My] husband was deeply impressed by a paper he read in a French scientific journal on sub-consciousness [*sic*]."[2] This unnamed article, she added, "gave the germ of the idea" that Stevenson afterwards developed into *Deacon Brodie, or the Double Life* (1880), a play he co-wrote with William Ernest Henley about the infamous eighteenth-century Scottish town councilor who led a secret nocturnal life of crime. Stevenson, we learn, then used this scientific inquiry again in his short story "Markheim" (1885), and, finally, "in a hectic fever following a hemorrhage of the lungs," it "culminated in the dream of Jekyll and Hyde."

The possible existence of this unidentified "paper on … subconsciousness" jars with Stevenson's own testimony, since the author denied using any medical theories or case studies as models. In an 1893 interview, a journalist from New Zealand asked Stevenson, "Had you heard of any actual case of double personality before you wrote the book?" Stevenson responded, "Never … after the book was published I heard of the case of 'Louis V.,' the man in the hospital at Rochefort. Mr. [Frederic W. H.] Myers sent it to me."[3] One is initially tempted to take Stevenson's account of the composition of *Jekyll and Hyde* more seriously than that of his spouse, not only because he was writing about these events at less remove (eight years after the composition of *Jekyll and Hyde* rather than twenty years) but also because Fanny's testimony elsewhere in her introduction is not entirely reliable.[4] Moreover, certain aspects of Stevenson's account can be corroborated. He did correspond with Myers shortly after the publication of *Jekyll and Hyde*, and Myers undoubtedly sent him a copy of his 1886 article "Multiplex Personality," which contains the case study of Louis V. mentioned by Stevenson.[5]

However, given the striking correspondences between *Jekyll and Hyde* and case studies published during the 1870s and 1880s, it seems unlikely that Stevenson's reply to the reporter was entirely honest. Richard Dury speculates that Stevenson "refuses to collaborate with the reporter because he does not wish to provide a single key to a story that is intended to remain enigmatic."[6] Critics including Dury, Jean-Pierre Naugrette, and Jacqueline Carroy have proposed various candidates for the "paper on subconsciousness," the most popular being a series of articles by Bordeaux physician Eugène Azam, published in *Revue Scientifique* between 1876 and 1879.[7]

In these essays, Azam introduced his famous patient Félida X., whom Ian Hacking identifies as "the first French double personality to be studied in depth."[8] Azam diagnosed Félida with "*double conscience*" or "*dédoublement de la personnalité*" (double consciousness or doubling of the personality).[9] Like other nineteenth-century cases of dual personality, Félida also presented hysterical symptoms such as sensory abnormalities, partial paralysis, tremors, spasms, and unexplained bleeding from the nose and mouth.[10] Her case was remarkable not only for its severity, but also because her two personalities were carefully documented over a long period of time: from 1858 well into the 1880s.[11] By the mid 1880s, Félida's fame had spread well beyond the boundaries of the French scientific community, to the extent that most English general readers had probably heard of her.[12]

While Azam's Félida is one probable source, I suspect that Stevenson had other prototypes in mind as well. In 1874, Jean-Martin Charcot's colleague Ernest Mesnet introduced Sergeant F., a soldier in the Franco-Prussian war who developed two distinct personalities after his left cerebral hemisphere was damaged by a gunshot wound.[13] The sergeant's case became famous when Thomas Huxley discussed it during his controversial lecture, "On the Hypothesis that Animals are Automata, and Its History" (1874), delivered as an address to the British Association for the Advancement of Science.[14] The sergeant's immoral behavior while in his "second state" strikingly resembles Hyde's misdeeds.

The case histories of both Félida and Sergeant F. were readily available to Stevenson in British periodicals. During the late 1870s, scientific journalist Richard Proctor (1837–88) summarized and translated these two patients' histories in a series of articles appearing in *Cornhill Magazine* (1860–1975). Julia Reid suggests the possible significance of Proctor's contribution in *Robert Louis Stevenson, Science, and the* Fin de Siècle (2006):

> It is perhaps likely that [Stevenson] read the scientific popularizer Richard Proctor's "Dual Consciousness" (1877), which appeared in the same *Cornhill Magazine* volume as an article by Stevenson, at a time when he was a regular contributor. Whether or not he knew of these developments, his novella certainly engages with the same issues, taking up for instance the question of moral responsibility raised by Proctor.[15]

While I believe that Reid is onto the right source material, she devotes less than a paragraph to the correspondences between Proctor's case studies and Stevenson's novella. Moreover, she does not mention Proctor's earlier *Cornhill* article "Have We Two Brains?" (1875), which contains a lengthy description of Sergeant F.'s case and raises further ethical questions about dual personality. Since Reid's study focuses primarily on Stevenson's role in Victorian evolutionary debates, however, it makes sense that she does not dwell at length on these sources.

In this chapter, I suggest that Proctor's articles "Dual Consciousness" and "Have We Two Brains?" were most likely Stevenson's main source of information on dual personality, based on numerous similarities between the aforementioned articles and *Jekyll and Hyde*. Further, I argue that Stevenson's novella does more than merely reflect the case studies upon which it is loosely based. Instead, *Jekyll and Hyde* creatively intervenes in late-Victorian debates about dual personality and its alleged cause, bilateral brain hemisphere asymmetry. This can be demonstrated through the novella's influence on later case studies of dual and multiple personalities. Myers's article "Multiplex Personality," for example, appeared several months after *Jekyll and Hyde*. One can easily trace parallels between Stevenson's Hyde and Myers's description of Louis V., as suggested by Stevenson's remarks to the reporter from New Zealand. Stanley Finger, meanwhile, contends that *Jekyll and Hyde* "could well have affected how some clinicians subsequently viewed their cases," suggesting that a case study written by Scottish psychiatrist Lewis Bruce in 1895 bears the mark of Stevenson's influence. This study featured a patient named H.P. who "seemed to have two distinct consciousnesses." Intriguingly, one of his personalities spoke Welsh and the other English. Moreover, one personality was right-handed while the other was left-handed, suggesting that each was dominated by a separate brain hemisphere.[16]

Stevenson was well qualified to intervene in scientific controversies due to his background, education, and personal relationships with scientists. The author came from a family of Scottish lighthouse engineers and studied engineering before turning to literature as a profession. He befriended scientific luminaries like engineer Fleeming Jenkin and educational psychologist James Sully, while corresponding sporadically with other important researchers, such as Myers and the renowned French alienist Pierre Janet.[17]

In *Jekyll and Hyde*, Stevenson amply demonstrates his familiarity with rhetorical conventions of scientific prose. The novella parodies the form of the case study in order to reveal the weaknesses of late-Victorian scientific narrative, specifically, how the linearity and emotional detachment of medical case studies obscures the baroque complexity of mental pathology. This parody of the case study takes place in the unlikely medium of the Gothic romance, the genre that critics most often associate with *Jekyll and Hyde*. Robert Mighall, for instance, sees *Jekyll and Hyde* as a model of "a new breed of Gothic fiction" that challenges realist notions of time, place, and context but still exhibits a marked "somatic and physiological character," while Patrick Brantlinger places Stevenson in the category of late-Victorian "Gothic romancers, whose stories always veer toward dreams and the subliminal reaches of the mind."[18] By mimicking the case study within a Gothic romance, Stevenson lays bare the limitations of scientific prose, particularly the difficulty of discussing complex moral and social realities in purely empirical terms. One might conceptualize Jekyll's "perennial war among [his] members," then, as the doctor's struggle to maintain scientific objectivity in the face of a terrifying subjective reality.[19]

Stevenson was no mere satirist of scientific conventions, however. He clearly endorsed particular scientific theories even as he implicitly mocked the rhetoric in which they were couched. The most likely origin of Stevenson's conception of dual personality as it appears in *Jekyll and Hyde* is the theory of the double brain, first developed by continental physiologists such as Austrian anatomist Franz Joseph Gall (1758–1828) and later imported to England by such physicians as Sir Henry Holland (1788–1873) and Arthur Ladbroke Wigan (1785–1847) during the first half of the nineteenth century. "[M]an is not truly one, but truly two," Jekyll relates, apparently supporting theories suggesting that each brain hemisphere might house a separate personality, or even a separate soul (48). Jekyll's lament that "these polar twins should be continuously struggling" likewise evokes contemporary scientific views that the left and right hemispheres not only differed in their abilities, but also occasionally exhibited contrasting desires and moral inclinations (49).

Victorian dual-brain theory suggested that one could possess a maximum of two distinct personalities, one stemming from each brain hemisphere. This point represents a crucial difference between dual personality and the better-known condition of multiple personality disorder (MPD), now called dissociative identity disorder, a condition made familiar to modern audiences in films like *The Three Faces of Eve* (1957) and *Sybil* (1976). The first case of *multiple* (as opposed to dual) personality was described in July 1885, when physicians Hippolyte Bourru and P. Burot began treating the aforementioned Louis V. (Vivet) and conducting a remarkable series of experiments. By exposing Vivet to various narcotic substances, magnets, and metals, the doctors elicited eight distinct personalities, each of which had a separate memory.

Vivet's case was first publicized in England the following year, in A. T. Myers's article "The Life-History of a Case of Double or Multiple Personality" and Frederic Myers's "Multiplex Personality," both of which appeared too late to have influenced *Jekyll and Hyde*.[20] Given Stevenson's knowledge of French and friendships with prominent scientists, it is possible that he heard of Vivet while writing *Jekyll and Hyde*, though it is impossible to prove one way or the other. Vivet's case was not the first time multiple personalities had been observed, merely the first time the condition had been so classified. For instance, Félida's doctors actually observed as many as five alternate personalities. However, her physicians recognized only two, since her additional alters did not fit existing clinical models.[21]

The etiology of MPD is radically different from dual personality, which was thought to result from a physical problem: specifically, bilateral brain hemisphere asymmetry. By contrast, dissociation in its various forms was linked to childhood trauma as early as 1889, in Pierre Janet's *Psychological Automatism*. This connection is also prominent in Sigmund Freud and Josef Breuer's *Studies on Hysteria* (1895), particularly in the famous case of Anna O., a hysteric whom some have sought to "rediagnose" with MPD.[22] This connection between MPD and childhood trauma persists even in today's biological determinist medical climate. In the fourth edition of the *Diagnostic and Statistical Manual of Mental Disorders* (1994), MPD was renamed "dissociative identity disorder," but its cause remained the same: long-forgotten child abuse.[23] By contrast, the dual personalities described in *Jekyll and Hyde* have a somatic origin, and are rooted in the biology of the brain rather than the vagaries of experience. The next two sections of this chapter will discuss the significance of dual-brain theory for the Victorians, particularly the connection between dual-brain theory and dual consciousness. While historians of science have extensively documented this relationship, literary critics have yet to appreciate its full significance for *Jekyll and Hyde.*

## *JEKYLL AND HYDE* AND DUAL-BRAIN THEORY

Historians of science were the first to recognize that dual-brain theory might be relevant to *Jekyll and Hyde.* In *Medicine, Mind, and the Double Brain* (1987), Anne Harrington writes, "One would have to argue … that Jekyll would tend to focus his personality in the civilized, rational left hemisphere, while Hyde would give vent to his criminal instincts from somewhere in the recesses of the uneducated, evolutionarily backward right hemisphere."[24] Finger makes a similar observation in *Minds Behind the Brain: A History of the Pioneers and their Discoveries* (2000), where he suggests that Dr. Jekyll is "largely the personification of the cultivated left hemisphere," while Hyde may be "the doctor's morally defective self, the growing personification of the 'primitive' right hemisphere."[25] Since Finger and Harrington are writing scientific histories, however, neither explores the narrative implications of Jekyll's duality.

Among literary critics, Elaine Showalter was the first to relate the dual-brain theory to *Jekyll and Hyde.* Showalter briefly elaborates upon Harrington's observations, arguing that "the dominant side of the brain represent[s] the dominant gender, and the other the repressed gender."[26] Showalter's invocation of the dual-brain theory is part of a queer critical reading of Stevenson's narrative that draws on Sigmund Freud and his contemporary Wilhelm Fliess, who "argued that all human beings were bisexual."[27] Yet, by orienting her discussion toward Freud and Fliess, Showalter's critique occludes the significant ways in which Stevenson's novella sought to give literary intelligibility to psychological theories of the 1870s and 1880s that differed markedly from psychoanalytic understandings of the drives. While late-nineteenth-century scientific thought displayed a marked tendency toward biological determinism, Freud was less concerned with finding physiological causes for psychological ills and more interested in formulating symbolic conceptions of mental dynamics (especially after 1895 or thereabouts).[28] Showalter's commentary is thus symptomatic of a larger body of Stevenson criticism in which Freud's later works overshadow any late-Victorian scientific sources from which Stevenson might have drawn.[29]

Most recently, two critical editions of *Jekyll and Hyde* have incorporated Victorian scientific writings about dual-brain theory into their notes and appendices, suggesting growing critical awareness of the theory's importance for Stevenson's novella. Katherine Linehan's 2003 Norton critical edition of *Jekyll and Hyde* includes segments of Myers's article, "Multiplex Personality," as an important reference for understanding Stevenson's story. These excerpts do suggest brain duality as a possible cause for dual and multiple personalities. For instance, Myers relates that Louis V.'s doctors could "inhibit" his left brain using magnets and stimulants. When they did so, Vivet became "not only left-handed but *sinister*." By contrast, inhibiting the patient's right brain caused his "higher qualities of character" to emerge, including "self-control," "modesty," and "the sense of duty – the qualities which man has developed as he has risen from the savage level."[30] But Myers's article post-dates Stevenson's novella, which Myers is known to have read and admired. It therefore seems likely, as Reid suggests, that "Myers was … much indebted to Stevenson" rather than the other way around.[31] Moreover, Stevenson likely drew inspiration from case studies written in the mid 1870s, when dual personalities were widely discussed, but "multiplex" personality had yet to be recognized as a clinical possibility.[32]

Martin Danahay's 2005 Broadview critical edition of *Jekyll and Hyde* gathers together even more material on dual-brain theory in an appendix devoted to Victorian psychology. The appendix includes portions of Henry Maudsley's article, "The Double Brain" (1889) and James Sully's "The Dream as Revelation" (1893), as well as Myers's "Multiplex Personality." All three works shed light on dual-brain theory and dual personality as they were understood in the years following the publication of *Jekyll and Hyde*.

One wonders, however, why Linehan and Danahay did not turn to earlier writers on dual-brain theory whose works might have influenced the composition of Stevenson's novella. In the following paragraphs, I will do just that. Examining mid-Victorian authors like Holland, Wigan, and Charles Édouard Brown-Séquard reveals the scientific roots of Stevenson's dual protagonist. Taken together, these writings suggest why dual-brain theory provoked a certain theological unease even as it purported to explain hitherto misunderstood medical phenomena. By heretically suggesting that man might have two souls (one per brain hemisphere), nineteenth-century writings on the double brain took on a Gothic tone that could be exploited and intensified in horror novels like *Jekyll and Hyde*.

Brain duality was one of the most controversial issues to emerge in early debates about cerebral localization. Efforts to localize cerebral functions began in the late eighteenth century with the advent of phrenology, although phrenologists lacked the surgical and experimental techniques necessary to support their claims.[33] As Harrington has related, phrenological pioneer Franz Joseph Gall promoted the idea that "each of the mental faculties existed in perfect symmetrical duplicate, with each pair localized in corresponding regions of the two hemispheres."[34] In other words, Gall taught that everyone has two perfectly formed brains, each of which can substitute for the other in cases of unilateral brain injury.

This claim immediately provoked controversy due to the potentially heretical suggestion that "each hemisphere at least was potentially capable of generating a 'soul' of its own, capable of independent will and consciousness."[35] Taking Gall's ideas to extremes, some of his contemporaries even argued that each phrenological faculty might generate a separate soul, so that "instead of *one* soul, phrenology gives us nearly forty."[36] This fanciful suggestion finds an echo in Jekyll's claim that "I hazard the guess that man will ultimately be known for a mere polity of multifarious, incongruous, and independent denizens" (48). This statement is a rare instance in which Jekyll speaks of multiplicity, rather than "the thorough and primitive duality of man" he elsewhere describes (49). One might even suggest that Stevenson here anticipates the phenomenon of multiplex or multiple personality. But I remain convinced that Stevenson was primarily interested in human duality, as this is the predominant theme running through "Henry Jekyll's Full Statement of the Case." Moreover, dual personality was the predominant clinical model available to Stevenson at the time that *Jekyll and Hyde* was composed, whereas multiple personality was only just beginning to be recognized as a possibility.

While Gall said that he never intended to challenge the teachings of the Catholic Church, his works were nonetheless placed on the Catholic *Index* of forbidden books, and he was refused a Christian burial.[37] Despite the religious opposition Gall encountered, his teachings were enormously influential, both in his native Austria and abroad. His most important legacy, Harrington suggests, was moving the imagined seat of the soul from the pineal gland (where René Descartes had placed it in the seventeenth century) to the cerebral hemispheres.[38] Gall had managed to convince the scientific community once and for all that the brain was the organ of the mind, a previously controversial suggestion.[39]

Gall's writing on hemispheric symmetry likewise had an enormous impact upon the next generation of brain researchers, including Henry

Holland and Arthur Ladbroke Wigan, who promoted Gall's dual-brain theory (with some modifications) in England during the 1840s. Wigan's *A New View of Insanity: The Duality of the Mind* (1844) introduced the idea that each individual had a moral imperative to keep both brain hemispheres balanced. Wigan argued that if one hemisphere became diseased, injured, or mad, the opposite (healthy) hemisphere "can still, up to a certain point, control the morbid volitions of its fellow."[40] In other words, the healthy hemisphere could serve as "a sentinel and security for the other," overriding the "erroneous judgments" of the diseased hemisphere.[41] Wigan's work is an early instance of how dual-brain theory was linked to madness, particularly dual consciousness of the sort Félida experienced.

Wigan's writing also had implications for educators, who were encouraged "to establish and confirm the power of concentrating the energies of both brains on the same subject at the same time; that is, to make both cerebra carry on the same train of thought together, as the object of moral discipline."[42] This idea of educating both hemispheres was revived in the 1870s by Brown-Séquard, who advocated ambidextrous education for all schoolchildren as a means of strengthening both halves of the brain. Since Proctor cites Brown-Séquard, Wigan and Holland extensively, it is likely that Stevenson was influenced by their theories of dual-brain function.

By the 1860s and 1870s, dual-brain theory had evolved considerably, thanks to a new breed of cerebral localization research taking hold in the scientific community. Scientists like Paul Broca, John Hughlings Jackson, and David Ferrier used evidence from autopsies and experiments on live animal brains to bolster their claims about hemispheric functioning. Their scientific methods were far more rigorous than Gall's technique of reading bumps on the skull to determine character traits. But their conclusions were likewise flawed, in part because these scientists favored evidence that supported their cultural biases. Steven J. Gould, for instance, has shown that Broca and his peers used experimental evidence "not to generate new theories but to illustrate a priori conclusions."[43] As a result, Victorian theories of hemispheric functioning tended to support existing race, gender, and class prejudices.

During the 1870s and 1880s, the idea of left-brain superiority became firmly entrenched, thanks to experimental evidence suggesting that "most or all of the higher 'intellectual' functions, presumably those associated with human beings alone, were housed exclusively in the left hemisphere."[44] Broca's localization of speech in the third frontal convolution of the left hemisphere certainly played a role in this development. Similarly, Hugo Liepmann's suggestion that the left hemisphere, "acting via the right hand, predominated in voluntary, purposeful movements" led many scientists to conclude that the human will was localized on the left side of the brain.[45] The left hemisphere thereafter became linked to civilization, rationality, and so-called "higher" cerebral functions such as language. Since these traits were traditionally associated with upper-class European males, it was supposed that the left hemisphere predominated within this demographic group. The dubious science of craniometry was called into service to verify this point, and batteries of skull measurements inevitably "proved" white male superiority based on brain shape and size.[46]

If the left hemisphere was civilized and rational, it seemed logical to many scientists that the right hemisphere should embody the opposite qualities: impulsivity, savagery, animality, and madness. Predictably, the right brain became associated with supposedly inferior groups such as women, non-whites, maniacs, and criminals, in whom it was supposed to predominate. Yet the right brain was also thought to have an important role in vital functions such as sleep, emotion, unconscious thought processes, and instinctual drives.[47] For instance, since Stevenson claimed that the idea for *Jekyll and Hyde* came to him in a dream, one might justifiably argue that the tale was a product of the author's right brain hemisphere.[48]

If one removes race, class, and gender bias from the picture, these perceptions match modern ideas about hemispheric function – up to a point. As Oliver Sacks has explained, "the left hemisphere is more sophisticated and specialized, a very late outgrowth of the primate, and especially the hominid, brain."[49] But although the left hemisphere evolved more recently, it is misleading to assume that the right hemisphere is truly "primitive." In fact, Sacks argues, the right brain controls "the crucial powers of recognizing reality which every living creature must have in order to survive," such as facial recognition, proprioception, and non-verbal communication.[50] But this was not fully understood in the mid nineteenth century, nor was it known that the right hemisphere plays an important role in certain visuo-spatial tasks.[51] Even today, neurologists know far less about the right hemisphere than the left.[52] The right hemisphere remains (relatively speaking) a dark and mysterious territory of the brain, just as it was during the Victorian era.[53]

As Harrington and Finger have suggested, the opposites embodied in the Jekyll/Hyde binary conform to late-Victorian ideas about the brain as a double organ. On the one hand, we have "Henry Jekyll, M.D., D.C.L., LL.D., F.R.S., &c." (13), "the very pink of the proprieties" (10). A respected and altruistic doctor, Jekyll appears a "large handsome," "well-made, smooth-faced man of fifty" (20, 19). On the other hand, Edward Hyde is "abnormal and misbegotten," exuding an ambiguous air of deformity (45). Unlike Jekyll, Hyde appears young and effeminate by virtue of his diminutive stature, dandyish tastes, and emotional lability, including an outburst of "hysteria" (45). Racially, Hyde is an ape-like evolutionary reversion sporting a "swart growth of hair" over the "dusky pallor" of his skin tone, in contrast to Jekyll's "white" skin (54). In other words, from a late-nineteenth-century perspective, Jekyll represents the pinnacle of evolution, while Hyde approaches its nadir.

In a strikingly predictable way, Jekyll exhibits left-hemisphere attributes (masculinity, whiteness, logic, intelligence, humanness), while Hyde embodies right-hemisphere traits (femininity, racial indeterminacy, madness, emotion, and animality). Such conflicting traits commonly surface in contemporary case studies of dual personality, a disease thought to be caused by imbalance of the brain hemispheres. As Proctor explained in 1875, "The promptings of evil and the voice of conscience resisting these promptings, present themselves as the operation of two brains, one less instructed and worse trained than the other."[54] According to the lateral model of brain asymmetry, the rational left brain would be roughly equivalent to "the voice of conscience" while the "worse-trained" right brain embodied the "promptings of evil" stemming from unchecked drives dating back to humanity's savage or animal past.

By housing left- and right-brain traits in separate characters, Stevenson's *Jekyll and Hyde* performs a fictional corpus callosotomy, splitting the nerve fibers that connect two brain hemispheres.[55] Intriguingly, Stevenson deemed surgical precision to be necessary for the creation of believable fictional characters. The author confessed shortly before his death that "psychical surgery is, I think, a common way of 'making character'; perhaps it is, indeed the only way … knife in hand, we must cut away and deduct the needless arborescence of [a character's] nature."[56]

Stevenson's figurative psychosurgery has dire moral and theological consequences for his dual protagonist. More than one Victorian scientist had been struck by the possibility that "so far as the brain represents it, the soul must be double."[57] Like these scientists, Stevenson explores the potentially heretical possibility that human beings are inherently double even in a healthy state. Jekyll explains that both he and Hyde existed before the discovery of the salt that enabled them to lead separate lives: "I learned to recognize the thorough and primitive duality of man; I saw that, of the two natures that contended in the field of my consciousness, even if I could rightly be said to be either, it was only because I was radically both" (49).

This passage locates humanity's essential doubleness in the distant evolutionary past. Both the epithet "primitive" and adverb "radically" (both of which mean, in this case, originary, at the root of) were used frequently in the emergent sciences of anthropology, geology, and the study of evolution during the late nineteenth century, often with racial overtones (denoting uncivilized, evolutionarily "backward" populations or "original" human populations from which modern races evolved). The proximity of these two words signals Stevenson's familiarity with debates following the publication of Charles Darwin's *Descent of Man* (1871). Moreover, the passage implies that duality is potentially consistent with mental health, and not necessarily a symptom of insanity.

If mental and spiritual duality can coexist with mental health, then Jekyll's madness must proceed from another cause: hemisphere imbalance. As Harrington has described, the 1870s and 1880s witnessed a revival of the idea that balanced hemispheres were necessary for mental health. Hemispheric imbalance was thought to be particularly dangerous when the development of the right hemisphere outpaced that of the left.[58] Victorian scientists argued that dual personality, along with other forms of insanity and criminality, resulted from a disproportionately large right brain overpowering the rational activities of the left brain. For example, in physiologist David Ferrier's article "The Brain of a Criminal Lunatic" (1882), the autopsy of the brain of a mentally ill woman who had murdered her children revealed an enlarged right brain and a "strikingly abnormal … atrophied" left brain missing a frontal lobe (where speech and rational functions were thought to be located; see Figure 1).[59]

Victorian scientists disagreed about whether such cases of hemisphere imbalance could be cured. While Henry Maudsley and other biological determinists denied this possibility, moral managers such as Wigan asserted that the mentally ill could realign their imbalanced hemispheres. Making an analogy between muscular exertion and brain activity, Wigan and (later) Brown-Séquard argued that using a given brain hemisphere promoted greater blood flow to that side of the brain. Various exercises were recommended for the purpose of strengthening one or both hemispheres. While Brown-Séquard recommended ambidextrous training for schoolchildren, adults could join "ambidextrous culture societies" where they would learn to perform two tasks at once, "such as playing the piano with one hand while writing a letter with the other." This influential

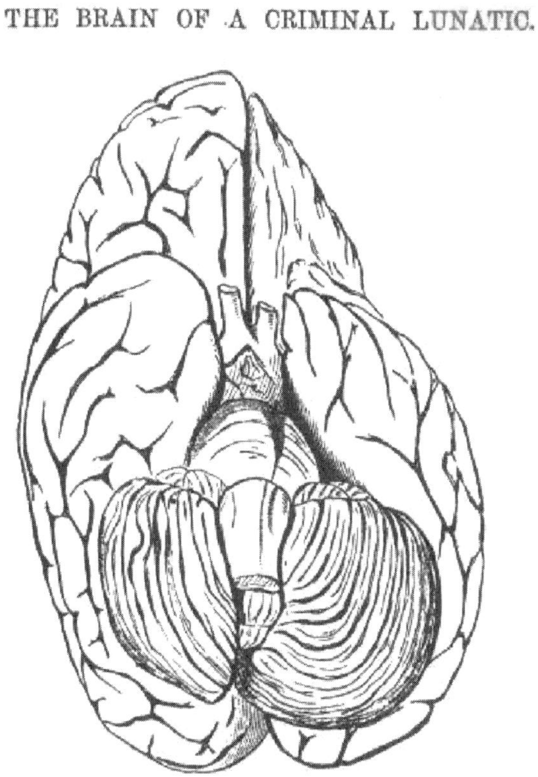


Figure 1 Diagram shows underside of brain. Notice that the left side of the brain has atrophied considerably.

movement lasted well into the twentieth century, despite vocal detractors who saw ambidextrous education as an extremist fad along the lines of "vegetarianism, hatlessness, or anti-vaccination."[60]

Stevenson's protagonist evidently lacks both the early training and the self-discipline needed to balance his hemispheres. Initially, the left-brained Jekyll overmasters his animalistic right-brain urges, necessitating the creation of the hedonistic secondary persona. This secondary persona, Hyde, begins as the weaker of the two:

> The evil side of my nature … was less robust and less developed than the good which I had just deposed. Again, in the course of my life, which had been, after all, nine tenths a life of effort, virtue and control, it had been much less exercised and much less exhausted. And hence, as I think, it came about that Edward Hyde was so much smaller, slighter and younger than Henry Jekyll. (51)

In Brown-Séquard's terms, the persona embodying Jekyll's "evil side" is smaller and less robust because under-used; Hyde represents the atrophied right hemisphere struggling to break free of the restraints imposed by the dominant left brain.

Predictably, Hyde increasingly predominates once he is unleashed with greater frequency. Jekyll's narration suggests an awareness of what generally happens when a hemisphere is used more often (i.e., a greater flow of blood to that side of the brain): "It had seemed to me of late as though the body of Edward Hyde had grown in stature, as though (when I wore that form) I were conscious of a more generous tide of blood; and I began to spy a danger that, if this were much prolonged, the balance of my nature might be permanently overthrown" (55). Jekyll's predictions are vividly realized by his eventual inability to "throw off" the body of Edward Hyde (55). From the perspective of Wigan or Brown-Séquard, Jekyll initially relies too heavily on his left hemisphere, and then shifts the balance too sharply toward the right. In each instance, he inadvertently creates the brain asymmetry that leads to his mental illness and criminality.

Physicians diagnosed hemisphere imbalance by looking for tell-tale bodily symptoms like left-handedness, signs which also surface in Jekyll and Hyde. Then as now, scientists thought that the left and right sides of the body were connected with opposite hemispheres of the brain.[61] Bodily left-sidedness (like left-handedness or left-leg predominance) signaled right-brain hemisphere dominance, which supposedly caused moral weakness and insanity.

Hyde's left-handed tendencies, demonstrated in his backward-sloping handwriting, thus present perhaps the strongest evidence of his right-brain dominance. Jekyll explains that "by sloping my own hand backward, I had supplied my double with a signature, [and] I thought I sat beyond the reach of fate" (53). Shortly after the publication of *Jekyll and Hyde*, Myers wrote to Stevenson that "Hyde's writing might look like Jekyll's done *with the left hand*."[62] If Hyde writes with his left hand while Jekyll uses his right, then this fact definitively links the two personalities with the right and left hemispheres of the brain, respectively.

## FÉLIDA X. AND SERGEANT F.

Stevenson probably came into contact with these theories of the double brain by reading particular case studies of dual personality, such as Richard Proctor's accounts of Félida X. and Sergeant F. in *Cornhill Magazine*. Although there were other popular examples of dual personality circulating during the 1870s and 1880s, particularly the famous case of Mary Reynolds, Proctor's essays would have been among the nearest to hand and most up-to-date available to Stevenson.[63] Moreover, Félida X. and Sergeant F. both exhibited symptoms that strikingly resemble Jekyll's split personality. Yet until now, no one has thoroughly investigated the correspondences between Proctor's widely circulated articles and Stevenson's novella.

This oversight seems surprising in light of Proctor's influence and extraordinary productivity during the mid-Victorian era. Proctor was a close friend of Grant Allen, whose writings form the subject of Chapter 3. He was also the founder and editor of *Knowledge: An Illustrated Magazine of Science*, not to mention the author of 57 books and over 160 popular science articles, on topics ranging from astronomy to "automatic chess playing."[64] Proctor was distinguished enough in his own time to merit an entry in *The Dictionary of National Biography*, but his influential career has been overlooked by scholars until quite recently.[65]

In his articles "Have We Two Brains?" (1875) and "Dual Consciousness" (1877), Proctor described Félida, a young hysteric who exhibited "a peculiar secondary state of mind" alternating with her "normal" personality. In her so-called normal state, Félida appeared "melancholy" and "very anxious about her bodily health."[66] In her second state, meanwhile, Félida "Woke up in quite another state, smiling gaily, speaking briskly, and trilling (fredonnant) over her work … and scarcely complained of any of the pains she had suffered so severely a few minutes before. She busied herself about the house, paid calls, and behaved like a healthy young girl of her age."[67] At first glance, Félida's disease apparently transforms her from a morose hypochondriac into a paragon of domestic virtue. Her change initially resembles Jekyll's, when the "younger, lighter, happier" Hyde succeeds his repressed and serious double (50).

Like Hyde's fateful shift, however, Félida's transformation had unfortunate moral consequences. In this respect, both of them conform to the prevailing clinical model of dual personality during the nineteenth century. Hacking relates that patients in their "normal state" were typically "docile, pious, and dull." During their *condition seconde*, meanwhile, these individuals lost their inhibitions and became vivacious, impulsive, and promiscuous.[68] True to form, Félida's "power of self-control … was manifestly weaker during her second condition."[69] While Proctor remained vague about Félida's misdeeds, her physician, Eugène Azam, explained that she became pregnant out of wedlock during her *condition seconde*: "A young man of 20 years of age knew Félida X. … the two young people had a great affection for one another, and were promised to one another in marriage. During her second condition, she abandoned herself to him and became pregnant. During her period of normal life, she ignored him."[70] Félida's sexual abandon during her second state does not obviously resemble Hyde's violent sprees. Stevenson's critics, however, have speculated that Jekyll's unnamed "irregularities" may be sexual transgressions. Gerard Manley Hopkins, for instance, memorably suggested that Hyde's violence substitutes for "something unsuitable for fiction."[71] Stephen Heath picks up on Hopkins's suggestion, arguing that Hyde's "random violence … has replaced a sexual drive."[72] Thus, Hyde's ambiguous misdeeds probably mirror Félida's sexual excesses. Stevenson's deliberate choice of a male rather than a female protagonist alters the gender politics of the situation, however, so that Hyde becomes a sexual predator ("*lusting* to inflict pain") rather than a fallen woman (59, emphasis added).

A potentially more crucial similarity between Félida X. and Jekyll is the manner in which they transform from one state into another. When Félida is changing states, she experiences "[s]harp pains attack[ing] both temples, and in a few moments she became unconscious. This lasted ten minutes."[73] Sergeant F. experienced similar symptoms: "The commencement of the abnormal state is ushered in by uneasiness and a sense of weight about the forehead" followed by "dullness and heaviness of the head."[74] When Jekyll takes the fateful powder, his symptoms resemble those mentioned above: "The most racking pangs succeeded: a grinding in the bones, deadly nausea, and a horror of the spirit that cannot be exceeded at the hour of birth or death. Then these agonies began swiftly to subside, and I came to myself as if out of a great sickness" (50). Later, when Jekyll changes into Hyde spontaneously, he feels "a qualm … a horrid nausea and the most deadly shuddering. These passed away, and left me faint" (58). On both occasions, there occurs a pleasant "change in the temper of my thoughts" after the physical agony subsides (58). Not only do the physical symptoms resemble Félida's, but the immediate psychological results of Jekyll's transformation into Hyde also mirror Félida's carefree abandon during her *condition seconde*. The increasing duration of their physical transformations likewise links Jekyll and Félida. As Dury notes, both "cases" exhibit "the growing and possibly complete dominance of the second personality-state."[75]

At least one contemporary reader of Stevenson's novella recognized the similarity between its divided protagonist and late-Victorian accounts of dual personality, particularly the symptoms that occur when a patient changes states. In a letter to Stevenson written in February 1886, Myers complimented the author on his portrayal of Jekyll's transformation. He suggested, however, that "there must have been a loss of consciousness" and that "the first time the loss of consciousness might last for some hours," followed by "more physical exhaustion."[76] Extended loss of consciousness and subsequent fatigue characterize Félida's and Sergeant F.'s transformations, but not Jekyll's. Myers's comments demonstrate his own familiarity with the case of Félida X. (whom he would later describe, along with Louis V., in "Multiplex Personality") and show that he urged Stevenson to make his novella conform more fully to contemporary scientific accounts of dual personality. That Myers appears to have understood Stevenson's story as a fictionalization of case studies like Félida's and Sergeant F.'s is powerful evidence that Stevenson actually used these models.

The case study of Sergeant F. was likewise described by Proctor in his *Cornhill* publications. Unlike Félida, whose disorder was naturally occurring, Sergeant F. developed two personalities as a result of a gunshot wound in the left brain hemisphere. In his "normal" state, Proctor wrote, "the ex-sergeant's health is perfect; he is intelligent and kindly, and performs satisfactorily the duties of a hospital attendant."[77] In his second state, however, the sergeant displayed animalistic, automatic qualities, along with impaired sensory impressions. At such times, he indifferently consumed "whatever is offered … asafetida, or vinegar, or quinine, as readily as water," and allowed pins to be run through his body without "causing the least indication of pain."[78]

Because of this insensitivity to pain and unpleasant tastes, Proctor characterized Sergeant F. in his second state as machine-like or bestial. Like Félida, Sergeant F. also exhibited diminished moral faculties after his transformation, as Proctor explained:

> In the unfortunate subjects of such abnormal conditions of the brain, the disturbance of the sensory and intellectual faculties is not unfrequently accompanied by a perturbation of the moral nature which may manifest itself in a most astonishing love of lying for its own sake. And in this respect, also, F's case is singularly instructive, for although in his normal state he is a perfectly honest man, in his abnormal condition he is an inveterate thief, stealing, and hiding away whatever he can lay hands on, with much dexterity, and with an absurd indifference as to whether the property is his own or not. Hoffman's terrible conception of the "Doppelt-gaenger" [*sic*] is realized by men in this state, who live two lives, in the one of which they may be guilty of the most criminal acts, while in the other they are eminently virtuous and respectable. Neither life knows anything of the other.[79]

Here we have the basic structuring concept of *Jekyll and Hyde*, whose protagonist oscillates between bourgeois and criminal personalities.

The above passage also highlights the similarities between dual personality and hysteria. In fact, dual personality was considered to be an unusual form of hysteria throughout the nineteenth century.[80] Both conditions were stigmatized as feminine, even though some patients were male.[81] Mark Micale relates how the hysteric, like the aforementioned cases of dual personality, had long been characterized in medical literature and popular lore as "a deceitful, adulterous, sexually treacherous creature: the hypererotic hysteric."[82] Sergeant F.'s "love of lying for its own sake" as well as Félida's sexual abandon, remind us of the same qualities in Hyde, suggesting that dual personalities could emasculate the male sufferer. The following section further elucidates how the social and sexual role reversals in *Jekyll and Hyde* contribute to Stevenson's polemical critique of late-Victorian medicine.

## *JEKYLL AND HYDE* AS GOTHIC CASE STUDY

Despite the compelling similarities between Stevenson's fictional *Strange Case* and Proctor's scientific case studies, Stevenson did more than merely fictionalize Proctor's material. Instead, he fused dual-brain theory and literary form, creating a Gothic story that parodies the supposedly objective format of the medical case study. One might view the Gothic as a "right-brain" genre due to its association with dreams and the unconscious, while the scientific case study appears to be a product of the rational left brain. If so, then *Jekyll and Hyde* suffers from a case of split personality much like Jekyll's own pathology.

To be sure, Stevenson did not confine his interest in dual personality to a single literary work. As Katherine Linehan, Masao Miyoshi, and others have observed, earlier works like *Deacon Brodie* and "Markheim" likewise grapple with the scientific and moral aspects of double lives.[83] But *Jekyll and Hyde* embraces contemporary scientific theories with greater thoroughness, containing an altogether more explicit address to medical inquiries than we find in these earlier works. Even the novella's full title, *Strange* Case *of Dr. Jekyll and Mr. Hyde*, explicitly invites us to read the work as a scientific case study.

At first glance, this novella's relationship to the case history is far from obvious because of its recognizably Gothic conventions: the nocturnal setting, the theme of monstrosity, and the embedded narratives (such as fragments, found manuscripts, letters, etc.). It is of course well known that *Jekyll and Hyde*, in exploiting these narrative features that date back at least to the time of William Thomas Beckford (1760–1844), belongs to the distinctive late-Victorian revival of the Gothic. As critics like Mighall have explained, the Gothic emphasis on psychological interiority and emotion might initially seem at odds with the rational aims of the scientific case study. Mighall suggests that Gothic "horror fiction has a generic obligation to evoke fear or suggest mystery," whereas "science … attempts to contain fear and offer a rational explanation for all phenomena."[84]

While the dual-brain theory Stevenson invokes was a Victorian scientific commonplace, the figure of the double is also a Gothic convention. One thinks, for instance, of James Hogg's *Confessions of a Justified Sinner* (1824), Charles Maturin's *Melmoth the Wanderer* (1820), or even Mary Shelley's *Frankenstein* (1818), where the monstrous creature enacts his creator's darkest desires. In some respects, *Jekyll and Hyde* strongly resembles *Frankenstein*, but differs in that it relocates the familiar Gothic topos of the ancestral home and transforms it into the space of the urban laboratory.

Despite its conformity to certain Gothic conventions, the novella succeeds in mimicking and critiquing the rational form of the case study through its seemingly dispassionate narrative voice. In Victorian scientific journals such as *Mind: A Quarterly Review of Psychology and Philosophy* (1876–present) or *Brain: A Journal of Neurology* (1878–present), the typical case study commences with an ostensibly objective third-person narrative, written by one or more scientists, with anecdotal and numerical evidence and illustrations appended toward the end of the piece. Similarly, Stevenson's novella starts out with a third-person narrator writing mainly from the point of view of Gabriel John Utterson, the lawyer whose rational detective work resembles the research of a scientist-author. The letters from Hastie Lanyon and Henry Jekyll at the end of the novella resemble the concrete data placed toward the end of the traditional nineteenth-century case study. In this instance, the fragmentary, epistolary structure of the Gothic novella neatly coheres with the traditional components of the case study.

Yet the apparent resemblance between Stevenson's romance and late-nineteenth-century medical case studies discloses a chink in the armor of late-Victorian scientific objectivity. Using anecdotal accounts of patients' histories, scientific authors constructed narratives suspiciously akin to fictitious productions. In Stevenson's well-known critiques of realist prose written during the early 1880s, he chipped away at distinctions that falsely differentiated fictional narratives from those supposedly based on objective truths: "in every biography with any salt of life … in every history where events and men, rather than ideas, are presented … the novelist will find many of his own methods most conspicuously and adroitly handled."[85] Stevenson did not add the case study to his examples of histories and biographies that use "the same technical manoeuvres" as works of fiction.[86] But elsewhere in his literary critical essays, Stevenson suggested that scientists and literary writers shared certain methodologies. He even denounced realist novelists for their "scientific thoroughness," arguing that realists willfully "dissect, with the most cutting logic, the complications of life, and of the human spirit."[87]

If realists are brutal vivisectors, then Stevenson and other authors of romances likewise operated according to scientific methodologies, principally mathematical rather than biological ones. Stevenson wrote of "a common organic law" uniting "the highest achievements in the art of words," further suggesting that this law might be geometric:[88] "The arts, like arithmetic and geometry, turn away their eyes from the gross, coloured and mobile nature at our feet, and regard instead a certain figmentary abstraction … a proposition of geometry is a fair and luminous parallel for a work of art."[89] While "[g]eography will tell us of a circle, a thing never seen in nature," the romance will likewise describe idealized events rather than facts: "those things which [the author] has only wished to do, than of those which he has done."[90] Stevenson, then, was no mere Luddite, but one who distinguished between different scientific methodologies available to authors of case studies, on the one hand, and authors of fiction, on the other hand. He implied that the detached rationality of the scientific author or realist novelist pales beside the idealized abstractions of the mathematician or the author of romances.

Stevenson's musings on literary form suggest what he stood to gain from writing a romance about a scientist who struggles to maintain his objectivity. The shift from omniscient narration at the beginning of the novella to the epistolary fragments that make up its concluding chapters parallels Jekyll's transition from objective physician to abject patient. Michel Foucault has argued that in the nineteenth century, "the locus in which [scientific] knowledge is formed" was not in the individual consciousness of a particular doctor, but rather "a generalized medical consciousness, diffused in space and time, open and mobile, linked to each individual existence."[91] This collective medical consciousness, which is suggested by the omniscient narration of the first two-thirds of the novella, finally crumbles when Lanyon's and Jekyll's individual accounts take over the narrative. The breakdown of objectivity dramatized in the novella's narrative form and in Jekyll's mental collapse demonstrates the ultimate triumph of the subjective Gothic romance over the objective case study or realist novel.

In Jekyll's last confession, the doctor hopelessly confuses the boundaries between objective observation and subjective experience. "Henry Jekyll's Full Statement of the Case" shifts between the first and third person, particularly in reference to Hyde: "He, I say – I cannot say, I" (59). Unsure even of his pronoun usage, Jekyll is finally divested of the armature of his scientific vocabulary and revealed as a lost soul. In his final letter, Henry Jekyll offers himself up as a case study to be observed by the penetrating gaze of fellow scientists. Through this role reversal, Stevenson questioned the power structures of the nineteenth-century medical establishment, particularly the inflexible divide between observer and observed, practitioner and patient.

In other words, Stevenson did not merely reproduce the typical form of the case study, which was generally dry, unemotional, and detached from the patient's suffering. Stevenson combined the basic structure of the case study with a tone and subject matter more appropriate to the Gothic, so that his novella itself suffers from a case of split personality. The logical, left-brain perspective of science combines with the primitive, emotional, right-brain perspective of the Gothic, demonstrating how Stevenson incorporated the polarities of the dual-brain theory into the literary form of his famous novella.

## STEVENSON'S INTERVENTION INTO VICTORIAN PSYCHOLOGY

In many respects, Stevenson's "case study" is a radical critique of the power structures of nineteenth-century medicine. By making his diseased protagonist a physician, Stevenson ironized the traditional relationship between objective scientist and abject patient. Moreover, his focus on the "strange case" of a male professional implicitly attacked the prerogatives of a patriarchal institution that for the most part ("Sergeant F." notwithstanding) treated female subjects. Showalter explains that "by the middle of the nineteenth century, records showed that women had become the majority of patients in public lunatic asylums."[92] This was particularly true of cases of dual personality, the overwhelming majority of which were women.[93] In a century that constructed madness as stereotypically feminine, male doctors appeared "not only as the possessors, but also as the dispensers, of reason, which they can at will mete out to – or take away from – others."[94] Stevenson's male doctor struggling to repress the feminine elements within himself destabilizes the paradigmatic power relations between male physician and female patient, as well as the identification of women with madness that Showalter so thoroughly elucidates.

Stevenson's protagonist, then, not only inverts the sexual hierarchy upon which clinical investigations were built, but also casts doubt upon the claims to objectivity upon which the late-Victorian case study had come to rest. As Jan Goldstein observes, nineteenth-century medical writing normally maintained a strict objectivity that prevented any identification with the patient: "The psychiatrist was not supposed to recognize the 'diseased' aspects of the patient in himself. A psychiatric diagnosis was in the nineteenth century something that a doctor gave to the patient as an 'other.' Not until the advent of psychoanalysis would subjectivity be valorized as an appropriate instrument of medical-scientific investigation."[95] In Stevenson's tale, by contrast, the doctor *is* the patient whose split subjectivity overwhelms his "Full Statement of the Case," making it difficult to tell whether it is Jekyll or Hyde who inscribes the "I" or the "he" of this document (47–62). The clinician's penetrating gaze has turned inward upon himself, fracturing him into two selves and upending the entrenched power structures of the medical establishment in the process.

Stevenson thus turned the tables upon the male professionals who sought to treat their patients dispassionately, revealing the limitations of the ways in which clinicians constructed their "case." Although Stevenson produced his critique of the case study ten years prior to Freud's development of psychoanalysis, he anticipated Freud's methods by exploring how a patient's split subjectivity might be incorporated into an ostensibly objective clinical report. As Hacking points out, psychological case studies after Freud began to challenge the idea of an objective medical collectivity: "Therapies, especially abreactive therapies, involve a relation between patient and therapist; what are we to make of the case reports when we know that the reporter is not a passive witness but an active agent in the case?"[96] Stevenson's novella dramatizes the same question in a unique and compelling manner, by housing "passive witness" and "active agent" in the same physical body. *Jekyll and Hyde* thus reads as a highly informed, daring critique of some of the nineteenth century's most revered assumptions about the diagnosis and classification of medical subjects, as well as a literary harbinger of crucial innovations in twentieth-century psychiatry.

But despite its inventive, forward-thinking aspects, *Jekyll and Hyde* takes a conservative stance with regard to cerebral localization theory. Stevenson transformed the polarities of the double brain into a tale of terror that shows the potentially disastrous consequences of hemispheric imbalance. Moreover, he reminded readers of the heretical implications of dual-brain theory, which suggested to Gall's critics and subsequent thinkers that each hemisphere might house a separate soul. Taken to extremes, one might view the brain's multiple functions as a conglomerate of "multifarious, incongruous, and independent denizens" whose presence makes the existence of a single soul highly dubious (48). In this context, it is important to keep in mind that *Jekyll and Hyde* was written at exactly the same time that dual personality fractured into multiple personality disorder with the case of Louis Vivet. The terrifying nature of this destabilizing transformation can be seen, if only obliquely, in Stevenson's novella about dual-brain function gone horribly awry.

CHAPTER 2

# *Bram Stoker's* Dracula *and cerebral automatism*

One of the most memorable scenes of horror in Bram Stoker's Gothic novel *Dracula* (1897) is the eponymous vampire villain's encounter with Lucy Westenra in the Whitby churchyard. Lucy, an upper-class English debutante whose nocturnal sleepwalking leads her straight into the arms of the vampire, explains what it feels like to be bitten in terrifyingly visceral terms:

> I have a vague memory of something long and dark with red eyes … and something very sweet and very bitter all around me at once; and then I seemed sinking into green water, and there was a singing in my ears, as I have heard there is to drowning men; and then everything seemed passing away from me; my soul seemed to go out from my body and float about in the air.[1]

Lucy remains surprisingly silent about the blood, fangs, crucifixes, and other items of vampire iconography widely associated with *Dracula*. The passage is also remarkably free of the sexual suggestiveness that usually surrounds Dracula's encounters with his female victims. For instance, nowhere does Lucy mention the penetration and exchange of bodily fluids that must have occurred during the abovementioned episode.

Instead, the horror of the passage resides in Lucy's detailed description of what it feels like to lose consciousness: to "sink," "drown," and lose one's individuality as everything external "pass[es] away." Although this loss of bodily control is itself terrifying, the culmination of Lucy's chilling experience occurs when her soul leaves her body. Later, once Dracula has transformed her into a vampire, Lucy becomes a shell of her former self, a "carnal and unspiritual … Thing which had taken Lucy's shape without her soul" (190). Lucy becomes a mere body that goes through the motions of survival without the guiding force of a soul or free will.

As fans of Stoker's *Dracula* well know, the novel's villain is so frightening in part because he lacks a soul, as evidenced by the fact that he casts no reflection in a mirror. He also possesses the ability to rob others of their souls when he transforms them into his vampiric minions. In other words, much more than human mortality is at stake in the vampire's depredations; free will and the possibility of an afterlife are also on the line. It is no exaggeration when lead vampire hunter (and brain specialist) Abraham Van Helsing describes his efforts to thwart Dracula as "this chess game which we play for the stake of human souls" (260).

This chapter elucidates how Dracula's swath of destruction parallels the progress of nineteenth-century science, particularly the transformation of physiological psychology from a discourse of the soul into a discourse of the brain. Like late-Victorian cerebral localizationists who argued that brain function was entirely determined by material factors (electrical stimulation, nutrition, etc.), Dracula threatens to transform his victims into human automata lacking souls or free will. J. Hillis Miller has memorably described how the Victorian era witnessed the disappearance of God from everyday life, thanks in part to rapid urbanization and emergent scientific developments such as Darwinian evolutionary theory.[2] In the realm of physiological psychology, this cultural sea change was reflected in the disappearance of the *soul*, which gradually vanished from mainstream scientific discourse in the wake of cerebral localization experiments and theories of cerebral automatism. In the seventeenth century, French philosopher René Descartes confidently located the soul in the pineal gland of the brain; by the late eighteenth century, however, phrenologists like Franz Joseph Gall were less definite about the soul's location, suggesting that it resided somewhere within the cerebral hemispheres.[3] By 1890, psychologists generally "ignor[ed] the question of the soul's seat," as William James relates.[4] These later researchers saw the soul as above or outside the domain of science, or, at the very least, as an entity that could not be studied using experimental methods.

How did this shift in priorities come about? In *Victorian Psychology and British Culture, 1850–1880* (2000), Rick Rylance explains how psychology transformed into a scientific discipline during the last few decades of the nineteenth century. During the first half of the century, psychology was conceived of primarily as the study of "the soul, or the spiritual principle in man."[5] Toward mid century, the more secular (but still incorporeal) concept of mind figured largely in psychological debate. The concept of mind, like that of the soul, indicated a thoroughgoing mind–body divide in psychological discourse, rather than the biological determinism that dominated the latter third of the century. The appeal of a soul- or mind-based psychology is obvious, as it preserves traditional Christian beliefs and permits considerable optimism about the possibility of spiritual reform. Mid-Victorian physicians, especially so-called "moral managers," emphasized man's ability to control his baser animal instincts by exerting the power of the soul or will. By contrast, late-Victorian biological determinists suggested that mental illness was a hereditary, incurable condition that manifested physically in the brain.

Although the "discourse of the soul" remained a potent influence within psychology throughout the nineteenth century, its impact waned as experimental physiology gained ground.[6] James and others have suggested that the 1870s were a crucial decade in establishing the physiological basis of psychology, due mainly to the work of cerebral localizationists who uncovered "the minute anatomy and the detailed physiology of the brain."[7] From the late 1860s onwards, localizationists like Gustav Fritsch, Eduard Hitzig, and David Ferrier experimented on live animals in order to connect specific aptitudes, emotions, and behaviors to distinct regions of the brain. Their crowning achievements included Ferrier's cranial maps, which "provided conclusive evidence of the cerebral localization of the major brain functions," Rylance explains.[8] The same period witnessed the professionalization of psychology as a discipline, including the founding of journals like *Mind: A Quarterly Review of Psychology and Philosophy* (1876–present) and *Brain: A Journal of Neurology* (1878–present) and the establishment of the first physiological laboratories. The net result of these developments was that the study of the mind or soul gave way to the measurement of physical phenomena occurring within the brain and nervous system.

The same decade sparked a controversial movement that became variously known as "the conscious automaton-theory" or the theory of human automatism.[9] These theories roughly correspond to the better-known philosophical concept of "epiphenomenalism," which is the view that mental events are caused by physical events in the brain, and that mental events have no causative power in and of themselves.[10] Drawing on the expanding prestige of late-Victorian physiology in general and cerebral localization in particular, so-called "automatists" like Thomas Henry Huxley, William Kingdon Clifford, and Shadworth Hodgson extended Descartes's idea that "brute animals are mere machines or automata" to human beings.[11] For instance, in his widely influential essay "On the Hypothesis That Animals Are Automata, and Its History" (1874), Huxley suggested that human "thought is a secretion of the brain" and that even the most complex series of ideas or emotions could be attributed to "reflex action" of the brain and nervous system.[12] In Huxley's view, intellect, reason, will, and other types of conscious thought could likely be localized in the frontal part of the brain.[13] When the frontal lobes of the brain sustained damage, a man could be reduced to "a senseless mechanism worked by molecular changes in his nervous system."[14]

Sergeant F., whose case is described in Chapter 1, served for Huxley as an excellent example of such a human machine. After a bullet fractured his left parietal bone, the sergeant periodically entered a *condition seconde* in which he ignored all sensory impressions except those of touch. During this second state, the sergeant was able to perform "marvelously complex operations … mechanically, and to all appearances without consciousness," including singing songs, dressing himself, and pantomiming a battle scene.[15] Sergeant F.'s case disturbingly suggested that human beings could function at a high level entirely without the aid of consciousness or willpower. Similarly, Dracula lacks a soul but possesses a "mighty brain" and "learning beyond compare," suggesting that intellectual powers need not have spiritual significance (263).

There can be no doubt that Huxley enjoyed playing the role of provocateur, deliberately shocking his audiences by reducing mind or consciousness to an "epiphenomenon" of the brain.[16] Yet he simultaneously tried to reassure his readers by suggesting that human conscious automatism did not necessarily betoken "fatalism, materialism, and atheism," pointing out that many devout Christians "have held more or less definitely the view that man is a conscious automaton."[17] Rather than reconciling materialism and religion, Huxley and other British researchers generally held that material and spiritual planes were parallel, so that neurological discoveries need not interfere with religion, and vice versa. Robert Young suggests that late-Victorian British researchers found this position liberating: "parallelism allowed [physiologists] the luxury of ontological agnosticism while they got on with their work."[18] Parallelism remains a popular philosophical position among twenty-first-century neurological researchers, as Joseph LeDoux explains. "Neuroscientists … typically start with the assumption that the materialist view of the mind-body problem is correct (that the mind is a product of the brain) and then try to understand how the brain makes the mind possible." However, "because philosophers and brain scientists are pursuing different concerns, progress in one field does not necessarily signal an advance or defeat in the other."[19] In other words, since God and the soul are immaterial, interaction between these spiritual entities and the material world need not violate the laws of physics.[20]

Some late-Victorian researchers hoped that the idea of parallelism would console readers distressed by the materialist implications of physiological research. Clifford, for instance, assured his audiences that physical and mental facts ran along parallel tracks, so that "there is no interference of one with the other."[21] But Clifford was also alive to the Gothic potential of cerebral automatism:

> The objection which many people feel to this doctrine is derived, I think, from the conception of such automata as are made by man … [i]f we consider, for example, a machine such as Frankenstein made and imagine ourselves to have been put together as that fearful machine was put together by a German [*sic*] student, the conception naturally strikes us with horror.[22]

Although Clifford tried to reassure readers that "our own case is not an analogous one," his Frankenstein metaphor suggests that Victorian audiences found the idea of cerebral automatism extremely frightening, on a par with the most creative Gothic monstrosities.

Indeed, cerebral localization and theories of human automatism provoked considerable backlash, especially within conservative factions of the scientific community. Commenting upon biological determinist views of brain function, American physician Oliver Wendell Holmes, Sr., lamented that "the destructive analysis of our new schoolmen threatens to distil away all we once called self-determination and free will, leaving only a *caput mortuum* of animal substance."[23] Holmes's remark demonstrates that he, for one, eschewed the parallelism favored by many of his colleagues. But he was certainly not alone in his belief that materialist physiological research violated revered spiritual principles.

The same could be said for eminent physiologist William Benjamin Carpenter, whose *Principles of Human Physiology* was, according to Huxley, the standard English work on the subject between 1842 and the early 1870s.[24] Carpenter famously developed the concept of "unconscious cerebration" in the 1850s; this idea was a variation of Thomas Laycock's theory of reflex action of the cerebrum, developed around 1838. Both theories describe how the brain, like the body, performs certain actions automatically, without conscious awareness or volition.[25] Our mental life, Carpenter implied, consists largely of spontaneous and involuntary nervous action. Although this theory would seem to support the concept of human automatism, Carpenter adamantly maintained that the Will (with a capital "W") could override the spontaneous activity of the brain when necessary.[26]

Like Holmes, Carpenter decried physiological research suggesting that the brain is no more than a sentient mechanism. Writing in 1875, a year after Huxley's famous lecture on automatism, Carpenter observed that the battle between "the partisans of Free Will, and the upholders of

Determinism ... has latterly been carried into the domain of Physiology, where the Determinist army has found a great accession of strength."[27] Carpenter felt that the "Determinist hypothesis" was insulting to human dignity and inconsistent with "the universal experience of mankind," which suggests to us that we have at least some control over our own destinies.[28] He further argued that

> [T]here is something in our *self*-consciousness – in our power, not only of picturing the external world to ourselves, but of reflecting upon our own mental states – and in our conviction of possessing a power of *choice* between two or more courses of action, whether mental or bodily, – which necessitates the conception of an Ego as something unconditioned by material states and physical forces.[29]

As one might perceive from the above remarks, late-Victorian opponents of biological determinism generally substituted terms such as "Will," "Ego," or "Consciousness" (often capitalized) for the soul, although they were essentially treading on theological ground.[30] Within the context of these psychological turf wars, Lucy's loss of consciousness (or Consciousness?) takes on a different, more terrible meaning.

Stoker's *Dracula* was part of the backlash that followed upon these changes in psychology. The vampire villain, who transforms people into soulless automata and ruthlessly experiments upon human beings, can be seen as a mad scientist of sorts: a stand-in for localizationists and automatists who argued that we are no more than the sum of our brain functions. Some of these researchers are mentioned by name in *Dracula*, including Ferrier, French neurologist Jean-Martin Charcot, and Claude Bernard's disciple, Sir John Scott Burdon-Sanderson, a notorious vivisector and professor of physiology at Oxford.

In its implied condemnation of cerebral localization research, *Dracula* exemplifies how Gothic novels in the tradition of *Frankenstein* represent "anti-science fiction, a form of apocalyptic fantasy verging on religious myth," as Patrick Brantlinger has convincingly argued.[31] But since Mary Shelley's Victor Frankenstein is a human being, it is at least possible to sympathize with him, despite his obvious flaws. By contrast, Stoker's mad scientist is demonic and inhuman, a fact that seemingly precludes all possibility of sympathy with the aims of localizationists. Neurologists are monsters, and they create soulless monstrosities. This is the reactionary message of Stoker's novel, which is far and away the most conservative work of fiction examined in this volume.

Perhaps unexpectedly, this profoundly antiscientific novel contains well-researched, up-to-date discussions of late-Victorian neurological theories. The pages that follow demonstrate how Stoker explicitly engaged with cerebral localization theories and the related idea of cerebral automatism throughout *Dracula*, despite his obvious distaste for the agnostic leanings of late-Victorian physiologists. To date, critics have largely overlooked *Dracula*'s stake in late-Victorian neurology. William Hughes explains that neglect of Ferrier, Charcot, and Burdon-Sanderson's work remains a significant gap in existing *Dracula* scholarship, even though other medical aspects of the text have been thoroughly explored.[32] For instance, Robert Mighall, Ernest Fontana, and Daniel Pick have convincingly linked *Dracula* to *fin-de-siècle* theories of evolutionary regression, criminology, and sexology, while others have explored blood transfusions and Victorian views on menstruation as relevant contexts for the novel.[33]

Coming closest to the domain of physiological psychology, recent work by Hughes, David Glover, and John Greenway has brought to light the novel's engagement with Carpenter's theory of unconscious cerebration.[34] While their work has helpfully elucidated previously unexplained aspects of the novel, particularly vampire hunter John Seward's two offhand references to "unconscious cerebration" (69, 237), it is important to remember that Carpenter represented a conservative voice in the debates surrounding cerebral localization and automatism, despite his qualified approval of Ferrier's work.[35] While Carpenter's observations on reflex actions of the brain seemed avant-garde in the 1850s, the eminent physiologist referred to himself (only half jokingly) as already "old" or "ossified" by the 1870s.[36] By the 1890s, most physiologists would have found quaint Carpenter's assertion that "the automatism of the Cerebrum is itself directed and controlled by some higher power."[37] Although Carpenter's physiological theories helped pave the way for scientists like Ferrier and Huxley, he clung to incorporeal entities like soul and will that physiologists largely ignored by the end of the century.

While Carpenter's relevance to *Dracula* has been the subject of much discussion, few scholars have addressed the novel's reference to Ferrier or the particulars of Renfield's brain surgery.[38] This chapter aims to redress this neglect by uncovering Stoker's ambitious research agenda while preparing *Dracula*. Some of his voluminous composition notes for the novel explicitly address cerebral localization, among other medical topics. I will also elucidate how Stoker's vampire villain resembles the scientists whose methods he imitates by "creeping into knowledge experimentally" (264). But first, it will be helpful to explain Stoker's conflicted attitude toward science, which resulted from the clash between his religious upbringing and his substantial scientific education.

## STOKER AND SCIENCE

Judging from *Dracula* alone, it would be difficult to say whether Stoker was a champion of scientific progress or a blinkered Luddite. On the one hand, Stoker's obvious enthusiasm for new-fangled technologies such as the phonograph, the typewriter, telegraphy, stenography, and even railway travel seems "nineteenth century up-to-date with a vengeance" (40).[39] The vampire hunters use these technologies to triumph over the older, supernatural powers represented in the figure of the evil Count. On the other hand, as I have suggested above, the novel waxes nostalgic about an obsolescent psychological discourse that privileges the soul over the brain. Dracula's eventual demise suggests the return to an older, hierarchical psychology in which the soul, will, and higher intellectual faculties triumph over automatic cerebral functions.

Stoker's conflicted view of science stems in part from his religiously orthodox upbringing. Stoker's biographers concur on this point. According to Barbara Belford, "[t]he Stokers were Protestant and attended the Church of Ireland regularly."[40] Paul Murray, meanwhile, describes Stoker's father, Abraham, as "a righteous man who was still counseling his children on morality when they were adults."[41] The son appears to have taken his father's lessons to heart; a Bible given to Stoker by his mother on his ninth birthday "is still extant and the underlining shows he read it with great care."[42] Despite these biographical details, it is hard to know how faithfully Stoker adhered to his early religious training in his later life. A possible clue lies in the muscular Christianity of *Dracula*'s vampire hunters, who wield crucifixes and refer to Dracula as "an arrow in the side of Him who died for man" (209).

Alongside Stoker's Protestant worldview, one must consider the author's impressive scientific literacy. Recent critics, particularly Carol Senf, have emphasized the author's extensive scientific education and mathematical acumen, although certain particulars of his training remain unknown.[43] Stoker probably took an undergraduate degree in a scientific field at Trinity College, Dublin around 1870, and a master's degree with honors in pure mathematics from the same institution around 1875. But exactly what subjects Stoker studied and when he graduated are issues contended by his biographers. Stoker claimed that he "had got University Honours in pure Mathematics," but Murray explains, "[t]wo students did graduate with honours in mathematics in the spring of 1870 but Stoker was not one of them."[44] Belford, meanwhile, writes that "[Stoker] graduated in 1871 with a degree in science and stayed on for a master's."[45]

Some of this confusion stems from the way Trinity structured its undergraduate education in the 1870s. J. V. Luce explains in his history of Trinity: "Up to 1855 foundation scholarships were awarded only in classics, but from then on students could also compete in 'science,' defined as philosophy, mathematics and physics."[46] This broad definition of the field of "science," particularly the inclusion of philosophy under that heading, reveals that an undergraduate "scientific" education was conceived of far more broadly in the nineteenth century than now. Because a Trinity degree in science at this time could incorporate classes in "geology, palaeontology, botany, and zoology" in addition to math and philosophy, it is difficult to know exactly what branch of scientific thought was Stoker's specialty, if any.[47]

How, then, do we discover what Stoker knew about medicine in general, and physiological psychology in particular? The author had close ties to several physicians, making it likely that he came into frequent contact with medical discourse. Senf explains, "Stoker came from a family that valued science and the rational inquiry associated with it as well as one that understood the importance of careful empirical research and of testing evidence rather than accepting faith or intuition."[48] In fact, three of Stoker's four brothers (Thornley, Richard, and George) became physicians. George, the youngest of the Stoker brothers, was consulting physician to the Lyceum Theatre in London, where Stoker served as manager from 1878 to 1898. George Stoker specialized in diseases of the throat which plagued the theater's actors. Meanwhile, Stoker's eldest brother, Sir William Thornley Stoker, was a distinguished physician who garnered titles and appointments, including a knighthood (conferred in 1895) and the presidency of the Royal College of Surgeons in Ireland.[49]

Bram Stoker also maintained social ties to well known medical men outside his family, especially Oscar Wilde's father, Sir William Wilde, a prominent eye and ear specialist who founded a Dublin hospital and was knighted in 1864.[50] Wilde became something of a surrogate father to Stoker, who visited the Wilde family salons both in Dublin and in London.[51] Sir William, like Stoker, maintained a healthy interest in both folklore and science, authoring a volume entitled *Irish Popular Superstitions* (1852) that included discussion of Irish vampire legends. Stoker included a nod to Sir William Wilde in his first novel, *The Snake's Pass* (1890), where Irish coachman Andy compliments a young man by comparing him to the distinguished physician.[52]

Although Stoker had access to modern medical theories by virtue of his education and ties to prominent physicians, his composition notes for *Dracula* reveal that he often turned to outdated sources in order to understand phenomena such as somnambulism, hypnotism, and catatonic states. For instance, Stoker consulted Robert Gray's *The Theory of Dreams* (1808), in which sleep is described as "a death whereby we live. A middle moderating point between life and death and is like death."[53] Such an uncanny state clearly appealed to a novelist intrigued by the occult.

*The Theory of Dreams* included accounts of famous or prophetic dreams and case studies of patients who lapsed in and out of death-like trance states, just like Dracula and his victims. Stoker transcribed parts of one such case study, that of "[a] woman of the name of Guasser, who was affected by a kind of catalepsy which attacked her twice a day, during which she sunk into a profound sleep, and was deprived of all internal and external sensation, her limbs grew hard and inflexible like stone … little pulse was discernible."[54] The parallels with *Dracula* evident here are even more obvious in a second account transcribed by Stoker:

> The case of Colonel Townshend, mentioned by Dr. Cheyne [author of *Cheyne's English Malady*, 1733] who was also very remarkable; he had for many years been affected with a nephritic complaint, and had the power of dying or expiring when he pleased, and afterward of coming to life again at pleasure … the pulse of the colonel gradually became insensible, no motion of the heart was perceptible, nor any symptom of life to be discerned, a mirror held to his mouth being not even soiled by his breath; he continued in this state near half an hour, and then gradually recovered.[55]

The reasons such cases interested Stoker are clear enough from *Dracula*, in which both the vampire and his victims enter and exit death-like trance states "at pleasure" or when commanded by the Count.

Like the strange cases described above, Dracula and his minions retain the appearance of life while in trance states, minus perceptible pulse or breathing. For instance, when Jonathan discovers Dracula "asleep" in his coffin, the unconscious Count looks "as if his youth had been half renewed," even though he exhibits "no pulse, no breath, no beating of the heart" (55, 53). Similarly, Lucy's death restores "part of her beauty," leading her mourners to question whether she has truly expired (147). Gray's volume also described remarkable instances of somnambulism in which persons apparently asleep perform complex actions demanding concentration, such as writing and "walking in their sleep over ridges and parapets," as Lucy very nearly does during her sleepwalking episode at Whitby.[56] Since Mina must traverse "the endless steps to the Abbey" in order to reach Lucy and bring her back home, it seems likely that Lucy took the same hazardous route to her seat on the cliff during her somnambulistic state (88).

Stoker's composition notes for *Dracula* also refer to Sarah Lee's *Anecdotes of the Habits and Instincts of Animals* (1852). Lee's entry on bats cited "the torpidity in which bats remain during the winter, in climates similar to that of England … like other animals which undergo the same suspension of powers, they have their histories of long imprisonment in places which seem inimical to life."[57] Hibernation must have fascinated Stoker because it is an extended period of nearly death-like sleep, in which the animal's vital functions slow down considerably. Dracula's and Lucy's periods of suspended animation likely owe something to Lee's descriptions of animal behavior.

It is not immediately obvious why Stoker turned to these early- and mid-nineteenth-century volumes on somnambulism, hibernation, and catalepsy when more contemporary accounts of such phenomena were readily available. For example, the cases Gray described closely resemble late-Victorian accounts of sleepwalking, automatic behaviors, and multiple personalities cited in the *Journal of the Society for Psychical Research* (1884–present) as well as in more mainstream periodicals. Possibly, Stoker chose these earlier scientific sources because he longed to return to an older psychology based on the soul. Gray's volume in particular expressed the idea that trance-like mental states have spiritual significance. He wrote, for instance, that "the continued activity of the mind, during the lethargy [of sleep or trance], is a just argument of its separate and independent existence."[58] By contrast, late-Victorian writers discussing trance states usually attributed them to some form of cerebral automatism in which the mind, will, or soul were either disengaged or altogether absent.

Stoker's preference for early-nineteenth-century psychology may also stem from his appreciation of John William Polidori's *The Vampyre* (1819), one of the most influential Gothic tales of the Romantic era. James Robinson Planché's *The Vampire; or, The Bride of the Isles* (1820), a loose adaptation of Polidori's novella, played a significant role in the early history of the Lyceum Theatre, and thus would have been familiar to Stoker. The famous "vampire trap" was devised for the first production of this play. This spring-controlled floor opening allowed an onstage villain to appear or disappear right before the audience's eyes.[59] As I have explained elsewhere, Polidori's *The Vampyre* and its various stage adaptations feature mesmeric vampires whose victims exhibit trance states and somnambulistic behaviors.[60] Polidori himself was a physician who wrote his medical dissertation on somnambulism at the University of Edinburgh in

1815. Stoker clearly adopted some somnambulistic traits as key features of Dracula and his victims, building on the foundation laid by *The Vampyre* and its stage interpretations. Needless to say, Polidori's fiction and scientific writing likewise drew upon an older psychology that privileged the soul (by then seated in the cerebral hemispheres) over lower functions. Like Bram Stoker's scientific sources for *Dracula*, then, the author's admiration for Polidori attests to his nostalgia for an obsolescent, soul-centered psychology that did not threaten traditional religious views.

## SCIENCE WITHOUT A SOUL

The most revealing detail of Stoker's composition notes for *Dracula*, however, lies in one brief, undated memorandum that touches upon the latest cerebral localization research, rather than the antiquated scientific theories Stoker generally preferred. This memo was written by Stoker's titled elder brother, Sir William Thornley Stoker (called "Thornley" by family and friends). Thornley was a surgeon at St. Patrick's Hospital in Dublin, otherwise known as Jonathan Swift's hospital "for fools and mad."[61] According to his obituary in the *British Medical Journal*, Thornley "took a special interest in surgery of the cerebro-spinal cavity" "on account of his connection to Dean Swift's foundation."[62] Beginning around 1890, he performed some of the first brain surgeries in Ireland apart from trepanning.[63] Using Ferrier's cortical maps, Thornley was able to save the life of a patient suffering from brain hemorrhage in the late 1880s, and to locate brain tumors and abscesses in patients he treated in the early 1890s. He described these avant-garde brain surgeries in several published case studies, in which he demonstrated how Ferrier's groundbreaking discoveries could be applied in clinical practice.[64]

But Thornley, who served as Inspector of Vivisection for Ireland from 1879 well into the twentieth century, was also acutely aware of Ferrier's infamous 1881 trial for violating the 1876 Anti-Vivisection Act. In the late 1880s, Thornley felt that the life-saving potential of Ferrier's discoveries amply justified his methods, and decried those "weak credulous, or mistaken" antivivisectionists who would "prefer that [a] man, formed in the image of his Maker, should die, rather than their feeble sentiment be offended by a painless experiment on an ape."[65] By the early twentieth century, however, Thornley had become a trenchant critic of certain types of vivisection, perhaps due to abuses he witnessed during his inspections of Irish laboratories. In 1906, Thornley made a statement before the second Royal Commission on Vivisection in which he denounced the use of animal experiments to illustrate medical lectures. He also stated that he generally opposed the use of live dogs and monkeys in experiments, because "they felt so much … the amount of terror that a dog felt even in being put under chloroform was very painful to witness."[66] Thornley's testimony provided valuable ammunition for animal rights activists, who considered him an ally in their fight for stronger antivivisection legislation.

Thornley's considerable expertise in physiological psychology – not to mention his insight into contemporary animal experimentation practices – made him an ideal resource for Bram Stoker while the author composed his most famous novel. The advice Thornley provided in his memorandum shaped some of the madhouse scenes in *Dracula*, particularly Renfield's unsuccessful trepanning operation. The memo features a diagram of the head and an indication that injuries to different parts of the brain would inhibit motor ability in specific ways (see Figure 2). Thornley wrote, "An injury to the side of the head … would produce symptoms in the opposite side of the body."[67] Referring to his diagram, he added: "If the pressure began at 1, the leg would be paralyzed; if at 2, the arm: if at 3, half the face."[68] It is no coincidence that these notations echo the descriptions of scientific experiments performed by localizationists like Ferrier, Fritsch, and Hitzig, who electrically stimulated specific portions of animals' motor cortexes and observed the resulting behaviors.

In fact, this injury and the procedure used to treat it strongly resemble a case Thornley wrote about some years earlier in his article "On a Case of Subcranial Haemorrhage Treated by Secondary Trephining" (1888). In this essay, Thornley described how he used Ferrier's cortical maps to save a patient's life. The patient sustained a brain injury near the fissure of Rolando that resulted in a hemorrhage over the right motor area, not to mention paralysis over the left side of the body (hemiplegia).[69] After Thornley trepanned and removed a blood clot and a tiny bone fragment, the man eventually recovered. Such a positive outcome was remarkable, given the low success rate of early brain surgeries.[70]

Except for its unsuccessful outcome, Renfield's injury and its treatment closely follow the scenario laid out in Thornley's memo and in this earlier case study. The unfortunate lunatic suffers from "a depressed fracture … haemorrhage… [and] blood clot" that necessitate "trephinning [*sic*] to remove the depressed bone."[71] Stoker utilized his brother's advice extensively to lend an aura of scientific authenticity to the death scene:

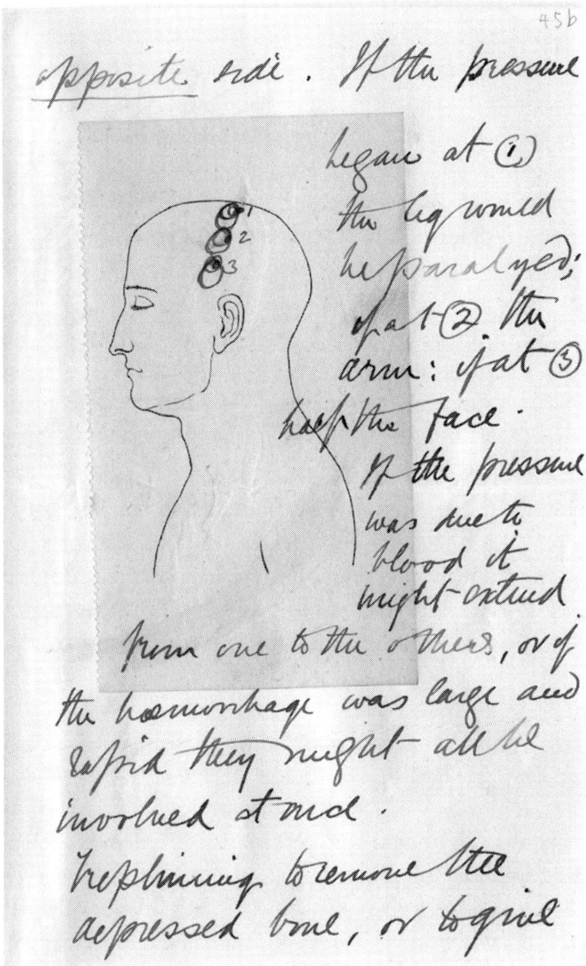

45b

opposite side. If the pressure began at (1) the leg would be paralysed; if at (2) the arm; if at (3) half the face. If the pressure was due to blood it might extend from one to the others, or if the haemorrhage was large and rapid they might all be involved at once. Trephining to remove the depressed bone, or to give

Figure 2 Sir William Thornley Stoker's "Memorandum on Head Injuries", written to his brother Bram Stoker, *c.*1890–6.

The real injury was depressed fracture of the skull, extending right up through the motor area. The Professor [Van Helsing] thought a moment and said: – "We must reduce the pressure and get back to normal conditions, as far as can be; the rapidity of the suffusion shows the terrible nature of his injury. The whole motor area seems affected. The suffusion of the brain will increase quickly, so we must trephine at once or it may be too late … we may most quickly and perfectly remove the blood clot; for it is evident that the haemorrhage is increasing. (242)

The scientific realism of this passage, and its nearly exact duplication of the details provided by Thornley, demonstrates Stoker's respect for his brother's medical expertise as well as his attempt to make his novel as scientifically accurate as possible. The scene, and the memo that inspired it, also attest to Ferrier's lurking presence in the background of Stoker's novel.

While Ferrier's cortical maps play a key role in Renfield's surgery, the famous neurologist is mentioned by name earlier in *Dracula*, when Seward ponders Renfield's curious mental state:

> Men sneered at vivisection, and yet look at its results today! Why not advance science in its most difficult and vital aspect – the knowledge of the brain? Had I even the secret of one such mind – did I hold the key to the fancy of even one lunatic – I might advance my own branch of science to a pitch compared with which Burdon-Sanderson's physiology or Ferrier's brain-knowledge would be as nothing. (71)

Seward's remarks suggest his unqualified admiration for both the methods and conclusions of Ferrier's research, not to mention the work of fellow vivisector (and sometime brain researcher) Burdon-Sanderson.[72] Because Seward is one of the novel's vampire-hunting protagonists, it would be tempting to assume that Stoker likewise approved of Ferrier's research agenda.

But Seward is hardly Stoker's mouthpiece in *Dracula*; in fact, he is a narrow-minded, bumbling character whose inability to put aside his scientific prejudices repeatedly precipitates disaster. As John Greenway has explained, Seward comes in for his fair share of narrative disapproval by virtue of his role as the novel's representative of "normal science." Using Thomas Kuhn's theory of scientific paradigm shifts, Greenway suggests that the intellectually adventurous Van Helsing pushes Seward to extend his definition of what constitutes normal science. Greenway likewise argues that Stoker favored Van Helsing's more inclusive ideas about what falls within the purview of scientific inquiry.[73] By initially rejecting outré ideas such as the possibility of telepathy and the existence of vampires, Seward stubbornly clings to the tenets of normal science at the expense of Dracula's intended victims. For instance, by overlooking Dracula's occult influence on Renfield, Seward fails to prevent the vampire from breaking into the madhouse and attacking Mina.[74]

Ferrier's localization research was certainly part of the late-Victorian scientific mainstream, which helps to explain why Seward embraces it. When Ferrier's groundbreaking work *The Functions of the Brain* appeared in 1876, it was very well received, and Ferrier was elected to be a Fellow of the Royal Society that same year. Like Stoker's brother Thornley, Ferrier was ultimately knighted for his accomplishments.[75] Subsequent researchers and historians of science have emphasized Ferrier's tremendous importance for future generations of scientists. Neurologist Charles Sherrington observed upon Ferrier's death in 1928: "Ferrier had been the main figure in proving the concept of cerebral localization, placing it at the centre of neurological interest, and providing the basis for a 'scientific phrenology.'"[76] Young, meanwhile, suggests the enduring impact of Ferrier's discoveries: "once the principle for cerebral localization was established, it provided a paradigm within which searching for centres [for motion, sensation, etc.] became, and to a large extent has remained, a part of normal science."[77]

Ferrier first became famous for a series of animal experiments conducted in the early 1870s at the West Riding Lunatic Asylum in Yorkshire.[78] Ferrier opened the skulls of monkeys, dogs, cats, rabbits, and other animals and stimulated specific areas of their brains with a mild alternating electric current. He then carefully observed the resulting movements and determined that stimulation of given brain regions reliably produced certain results. For instance, stimulation of region 5 of a macaque monkey's cerebrum caused "extension forward of the opposite arm and hand, as if to reach or touch something in front" (see Figure 3).[79] Stimulation of region 13, meanwhile, resulted in contracture of the pupils, and "closure of the eyelids as if under the stimulus of a strong light."[80]

Ferrier's detailed descriptions of these animals' behaviors reveal that he was not interested in reflex actions. Instead, he aimed to provoke volitional, directed movements that an animal would want to make under certain circumstances. Since the late eighteenth century, scientists had been aware that electrical stimulation could produce twitching and other reflex actions in corpses and living animals. By contrast, Ferrier stated that the movements of his experimental animals "are purposive or expressional in character, and such as we should, from psychological analysis, attribute to ideation or volition if we saw them performed by others."[81] In other words, Ferrier was not just making animals move, he was making them *want* to move. As Laura Otis has explained, Ferrier's experiments alarmed Victorian scientists and laypeople alike by demonstrating that an electrical stimulus could be substituted for volition. If such a thing were possible, then "there was nothing sacred about the human will, not even human consciousness."[82]

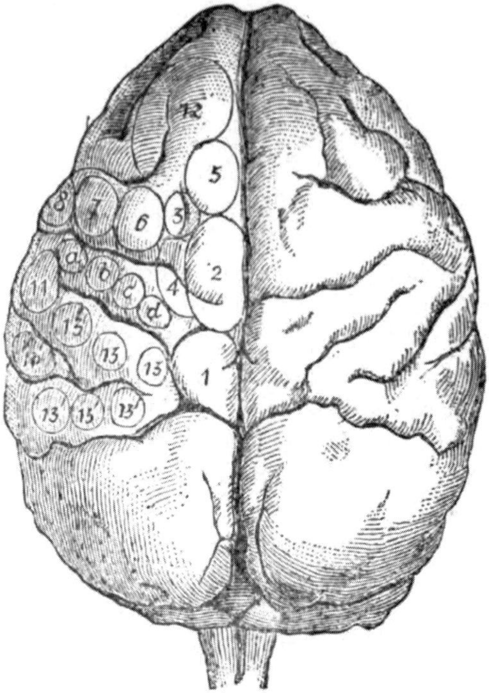

Figure 3 "Upper surface of the hemispheres of a monkey" from Sir David Ferrier, *Functions of the Brain.*

Ferrier's experiments also unsettled Victorians because they irrefutably demonstrated the similarity between human beings and other animals, whose brains were structured along similar lines. In fact, because human and monkey brains resemble each other so closely, Ferrier was able to extrapolate the results of his experiments on macaques to create the aforementioned cranial maps of the human cerebrum.[83] Ferrier's work provocatively elided the boundaries between human and animal in the same way as Darwin's *Descent of Man* (1871). But unlike Darwin, Ferrier made no attempt to assuage his audience's fears. In his 1878 volume, *The Localization of Cerebral Disease*, Ferrier ominously intoned, "It will be my endeavor to show you that what is true of the monkey is strictly true also of man."[84]

By the early 1880s, Ferrier's research was widely respected within the scientific community, but relatively unknown to the British public. This situation was about to change dramatically due to a series of events in

1881. That year, at the International Medical Conference in London, Ferrier exhibited several monkeys with damaged motor cortexes in order to demonstrate the principles of cerebral localization. He wished to refute the arguments of fellow presenter Friedrich Goltz, a German physiologist who believed that the cerebral cortex functioned holistically and contained no specialized organs. To prove his point, Goltz displayed dogs with cortical injuries who were able to function relatively well despite the cerebral damage they had sustained. Ferrier and his supporters later argued that the injuries sustained by Goltz's dogs were not as extensive as the German scientist had claimed, which would explain the dogs' relatively high functioning.[85]

To demonstrate the principle of localization, Ferrier displayed a hemiplegic monkey whose left motor cortex had been damaged as part of an experimental surgery. The monkey could only use its left hand and showed no ability to move the limbs on its right side. Upon seeing this unfortunate creature, Charcot famously exclaimed "It's a patient!" because the monkey reminded him of some of his charges at the Salpêtrière.[86] Ferrier also displayed a deaf monkey whose auditory cortex had been damaged. The two monkeys were afterwards sacrificed so that scientists could verify that their brains had been damaged in the manner Ferrier described.[87] Ferrier's demonstration convinced many scientists of his working hypothesis that "certain parts of the brain have determinate functions."[88]

Upon hearing of Ferrier's exhibition, animal rights activists were not so impressed. The members of the Victoria Street Society, an antivivisection group headed by Frances Power Cobbe, were appalled to learn that Ferrier had kept his monkeys alive for days, weeks, even months following their cranial surgeries in order to witness the long-term effects of cerebral damage. They also noticed that Ferrier had not precisely followed the procedures outlined in the 1876 Anti-Vivisection Act. In order to experiment on animals once anesthesia had worn off, a scientist was required to apply for a Certificate B, but Ferrier had not done so.[89]

Seizing their opportunity, Cobbe and her associates hauled Ferrier into court for violating the 1876 Act. The ensuing trial whipped up controversy surrounding vivisection in the popular press. Scientists assured the public that animal experimentation would result in discoveries that saved human lives, pointing out that Ferrier's cranial maps had already led to successful surgical treatment of brain tumors, blood clots, and skull fractures. Meanwhile, antivivisectionists painted their opponents as sadists who enjoyed inflicting pain on helpless animals. In her suggestively titled book, *The Modern Rack: Papers on Vivisection* (1889), Cobbe condemned scientists who "think of a brain which holds all the wondrous instincts and reasonings of the dog and the ape as a lump of grey matter to be scooped out and broken up … to note what happens after its mutilation."[90] She particularly deplored the treatment of "Professor Ferrier's monkeys," whose initial brain surgeries were followed by "hours, days, and weeks of misery before the victim dies."[91]

Scientific luminaries including Charcot, John Hughlings Jackson, and many others rallied to Ferrier's defense at his trial. From a scientific perspective, Ferrier's animal experiments made good sense – after all, one could not study the long-term effects of brain injury without keeping an animal alive after surgery. Moreover, the use of anesthetic could have compromised Ferrier's experimental results. Ferrier explained in *Functions of the Brain*, "[t]hat which will cause intense and indefinite action in an animal non-narcotized, will excite only moderate and definite action in an animal sufficiently narcotized to abolish all sense of pain, and no effect at all on an animal deeply anaesthetized."[92] None of these explanations were truly necessary, however, once it was discovered that Ferrier's colleague, physiologist Gerald Francis Yeo, had performed the experiments on monkey brains and successfully applied for a Certificate B. The charges against Ferrier were dismissed, leaving antivivisectionists fuming over the "entire futility of the present [Anti-vivisection] Act and the hopelessness of any proceedings under it."[93]

In fact, the antivivisection movement never entirely recovered from this highly publicized blow to its credibility. Increasingly, animal rights activists like Cobbe found themselves outnumbered and politically outmaneuvered by the "new priesthood" of scientists. As Coral Lansbury explains of the late-Victorian and Edwardian periods, "If faith in God and his Christian churches was waning throughout society, a fervent secular belief in science was rapidly taking its place."[94] Vivisection became an established part of the "secular religion" that predominated in *fin-de-siècle* England. Today, of course, animal experimentation remains a cornerstone of scientific practice, despite grassroots and academic animal rights movements that arise from time to time.[95]

Ferrier's infamous 1881 trial lingered in public memory long after he was officially acquitted of animal cruelty charges. To be sure, Ferrier's opponents were angered by his methods, particularly the fact that some of his "brain-mangling" animal experiments were conducted after anesthesia had worn off.[96] But, as Otis has suggested, Victorians were equally disturbed by the philosophical implications of Ferrier's work, particularly the way his research elided the boundaries between humans and animals and apparently undermined free will. Otis argues that the trial inspired Gothic fictions such as Wilkie Collins's *Heart and Science* (1883) and H. G. Wells's *The Island of Doctor Moreau* (1896), both of which implicitly critique aspects of Ferrier's work.[97] Due to uncanny parallels between Stoker's vampire villain and the famous neurologist, *Dracula* can likewise be read as a fictional condemnation of Ferrier's techniques and conclusions, and a particularly damning one at that.

Ferrier's research clashed with the traditional, religious worldview espoused in *Dracula*, in which human souls, willpower, and intellect ultimately triumph over vampiric automata. The physiologist served as a convenient target for Stoker because of his thoroughgoing materialism. Ferrier took a more agnostic stance than the researchers who preceded him, such as Fritsch and Hitzig, whose 1870 experiments demonstrated the electrical excitability of the brain and pinpointed the location of the motor cortex in dogs.[98] These trailblazing German researchers were surprisingly conservative in their views about the soul. They believed that the brain was the material agent of the immaterial soul, and cautiously concluded that "the origin of at least some function of the soul is bound up with circumscribed parts of the brain."[99]

By contrast, Ferrier had no interest in the soul, viewing it as outside the scope of his research. In fact, one of Ferrier's goals was to prove experimentally that *all* intellectual functions are strictly material, *especially* those typically ascribed to the soul or will. He argued in his groundbreaking work, *The Functions of the Brain*, for instance:

> We have every reason for believing that there is, in company with all our mental processes, *an unbroken material succession*. From the ingress of a sensation, to the outgoing responses in action, the mental succession is not for an instant dissevered from a physical succession … mental operations in the last analysis must be merely the subjective side of sensory and motor substrata.[100]

In the same paragraph, Ferrier derided researchers who suggested a "rupture of nervous continuity" due to the immaterial agencies of the mind, soul, will, etc. "It would be incompatible with everything we know of cerebral action," he wrote, "to suppose that the physical chain [of brain functions] ends abruptly in a physical void, occupied by an immaterial substance."[101] Although Ferrier left room for parallelism – the possibility that spiritual entities exist but do not interfere with brain function – his tone suggested his disdainful opinion of religion and spirituality more generally. Rather than soften his opinions to appease his readers, Ferrier stated unequivocally that "mental operations" are subservient to their physical causes.

Perhaps Ferrier felt justified in ruffling a few feathers, based on the magnitude of his accomplishments. Today, no one disputes the value of his contributions to science. But Ferrier's groundbreaking discoveries came at a significant psychological cost to the Victorian public. Stoker's *Dracula* provides ample evidence of this psychological toll in the frightful swath of destruction left by its eponymous vampire villain.

For Dracula replicates Ferrier's experiments in perverse and distressing ways. He first "narcotises" his victims via hypnosis, dulling the pain of his initial penetration. Victims later wake to pain and weakness due to their excessive blood loss, similar to how Ferrier's animals regained consciousness after their anesthesia had worn off. Dracula then electrically stimulates victims' brains using mental telepathy. In the 1880s, psychical researchers like Frederic Myers, Sir Oliver Lodge, and Sir William Barrett felt that they had proven the existence of telepathy, and speculated that telepathic communication was electrical in nature. For instance, physicists Lodge and Barrett published articles in the early 1880s in which they compared mental telepathy between brains to electrical induction.[102] When Mina and Renfield communicate telepathically with Dracula, then, their brains are reacting to the vampire's powerful electric influence. Finally, Dracula sacrifices his victims to feed his bloodlust, while Ferrier's animals were sacrificed for science.

Van Helsing himself describes Dracula as a first-rate scientist. During Dracula's lifetime, Van Helsing says, "he was no common man; for in that time, and for centuries after, he was spoken of as the cleverest and the most cunning, as well as the bravest of the sons of the 'land beyond the forest.' That mighty brain and iron resolution went with him to his grave, and are even now arrayed against us" (212). Members of the Dracula family, Van Helsing relates, studied at "the Scholomance … where the devil claims the tenth scholar as his due" (212). Pupils learned "the secrets of nature, the language of animals, and magic spells," in addition to alchemy, which "was the highest development of the science knowledge of [Dracula's] time" (263).[103] Dracula's background in alchemy and magic links him to the mad scientist tradition that began with sixteenth-century Faust legends.[104] Dracula even bears the physical stigmata that Victorians associated with scientific genius, including a "lofty domed forehead" bespeaking highly evolved intellectual faculties (23).[105]

Van Helsing also observes that Dracula's methods resemble those of late-Victorian physiologists. The Count is "experimenting, and doing it well … of late, this monster has been creeping into knowledge experimentally" (263–4). The Dutch physician cites Dracula's "experiments" on Renfield as one key example. Like Ferrier, Dracula uses a wide range of creatures as his experimental subjects: not just human beings, but also wolves, rats, bats, and other lower life forms. When convenient, Dracula even poses as a gentleman scientist of sorts. When the Count brings coffins of consecrated earth to England, he describes them on an invoice as "fifty cases of common earth, to be used for experimental purposes" (200).

Dracula's experiments on human beings flout a late-Victorian scientific taboo that Ferrier certainly never transgressed. Yet other physiologists were willing to cross this line. Take, for instance, Jean-Martin Charcot, who surfaces in *Dracula* when Van Helsing and Seward praise his scientific use of hypnotism and lament his recent passing (Charcot died in 1893, when the novel's events supposedly take place). Stoker probably met the renowned scientist, whom he proudly lists as a visitor to the Lyceum in *Personal Reminiscences of Henry Irving* (1906).[106] Yet whether or not Stoker was aware of it, Charcot's experimental techniques were arguably more diabolical than Ferrier's.

Whereas Ferrier was justly famous for localizing many cerebral functions, Charcot is today best remembered for labeling and classifying neurological diseases, and for his work on hysteria. An animal lover who owned a pet monkey named Rosalie, Charcot proudly displayed a sign at the entrance to his office stating: "You will find no dog laboratory here."[107] While he spared dogs and monkeys the fate they would have suffered elsewhere, Charcot had no qualms about experimenting on *people*, the very crime for which Dracula is so reviled. Charcot was known for his degrading examinations of nude patients and his exhibitions of particularly interesting cases at his Tuesday lectures. Some of his more vocal critics, such as Léon Daudet, accused him of being "abusive and insensitive" during such demonstrations.[108] His students and followers were sometimes called the *Charcoterie*, which was, as Stanley Finger explains, "a play on the French word *charcuterie*, meaning a butcher shop for pork products."[109]

English scientists were sometimes accused of similar crimes by antivivisectionists, who feared that poor people in hospitals might fall victim to enterprising or callous physicians. Anna Kingsford, a physician and antivivisectionist, complained that doctors performed some surgical operations on poor patients without the use of anesthetic: "Paupers are thus classed with animals as fitting subjects for painful experiment, and no regard is shown to the feeling of either."[110] Women were particularly sensitive to the plight of experimental animals and pauper patients, Lansbury suggests, because they themselves felt victimized by the invasive practices of Victorian gynecology. This may help to explain why "women were to be the strength of the antivivisection movement … every dog or cat strapped down for the vivisector's knife, reminded them of their own condition in society."[111] In this context, it is worth noting that most of Dracula's victims are women, with the exception of Renfield (and would-be victim Jonathan Harker). Like Charcot and many English physicians, Dracula is indifferent to the plight of women, using them for his own ends with no more compunction than he feels for his animal minions.

Yet perhaps we should not be so quick to condemn Dracula for experimenting on humans, since he is arguably of a different species. Mina says of Dracula, "this Thing is not human – not even beast," while Van Helsing observes that the Count "is of cunning more than mortal" (202, 209). Although the vampire hunters often cite Dracula's atavistic qualities, implying his evolutionary backwardness, they elsewhere suggest that Dracula belongs to a superior race, given his immortality, fecundity, and superior intelligence. The latter possibility is more disturbing, since it suggests the potential for species takeover (with vampires replacing humans as the dominant species on Earth). Van Helsing is well aware of this possibility. He warns Seward that Dracula "would be yet – he may be yet if we fail – the fatherer or furtherer of a new order of beings" (263).

If Dracula is a member of a superior race, then his experiments on human beings scarcely differ from vivisectors' use of animals for scientific investigations. By conducting cruel experiments on helpless members of an inferior species, Dracula resembles physiologists who justified their treatment of animals based on human intellectual superiority. Ironically, these scientists – many of whom were agnostics – also cited humans' possession of immortal souls as a quality that sets us apart from lower animals and gives us the right to do as we please with them.

Cobbe skewered such attitudes in a brief morality play, "Science in Excelsis" (1889), in which angels descend from heaven to perform cruel experiments on human physiologists. The angels turn the physiologists' own words against them, playing on the idea that animals are automata (yet another argument frequently used in defense of vivisection): "Grave doubts may be entertained as to whether Men are anything more than Automata; but, granting they have some dim feelings of pain and pleasure, it would surely be absurd for a moment to put their sensations in competition with the noble thirst for knowledge now stirring in the Angelic mind."[112] The angels then state their intention of replicating experiments from Burdon-Sanderson's "Handbook of the Physiological Laboratory" [*sic*], using physiologists rather than frogs and rabbits.[113] The angels propose to begin with the Englishmen, then move on to the more devious French and German investigators: "When we have sawn through their backbones, and irritated the stumps of their nerves, and rubbed caustic on their eyes, and made a few other interesting demonstrations, we shall be in a better mood to bake, and skin, and try many curious experiments with the rest."[114] Cobbe's descriptions are no exaggeration of the contents of late-Victorian physiological manuals, some of which did, indeed, provide instructions for baking rabbits alive. Moreover, Burdon-Sanderson's *Handbook* and similar treatises did not always insist on the use of anesthetic while performing such experiments.

If angels might plausibly descend from heaven to experiment on inferior human beings, couldn't a highly intelligent vampire do the same? *Dracula* turns on a similar premise to Cobbe's "Science in Excelsis," but offers up a more appropriate and terrifying villain: a malevolent, superior being who treats humans like guinea pigs. Given that Stoker's brother was an Inspector of Vivisection and an outspoken critic of certain types of animal experimentation, one cannot overlook the novel's references to Charcot, Ferrier, and Burdon-Sanderson, not to mention Seward's brief mention of vivisection. These hints, combined with Dracula's experimental acumen, suggest that the Count is an amalgam of the famous scientists who tormented animals in the service of a soulless psychology. No wonder Stoker wrote in his first novel, "For real cold-blooded horror, commend me to your men of science."[115]

## PSYCHICAL RESEARCH

If the chapter ended here, we could comfortably label Bram Stoker an antivivisectionist Luddite and move on. But Stoker's feelings toward science were more complicated, since he did, in fact, advocate a brand of research that exceeded normal science in its willingness to explore occult subjects like telepathy and the existence of immortal souls. In *Dracula*, Van Helsing serves as the representative of this more inclusive science. We learn about Van Helsing's beliefs when he chides his narrow-minded former pupil, Seward, for overlooking occult explanations of Lucy's death: "Ah, it is the fault of our science that it wants to explain all; and if it explain not, then it says there is nothing to explain ... I suppose now you do not believe in corporeal transference. No? Nor in materialization. No? Nor in astral bodies. No? Nor in the reading of thought. No? Nor in hypnotism" (171). Seward breaks in at this juncture, saying, "Charcot has proved that pretty well" (171). While Seward is willing to accept hypnotism because it has been incorporated into the normal science of the period, he is decidedly more hostile to the other phenomena Van Helsing mentions, all of which were topics of interest within late-Victorian psychical research.[116] Nina Auerbach and David Skal helpfully gloss Van Helsing's terminology, which resembles that used by the Society for Psychical Research (SPR): corporeal transference ("movement of objects by thought"); materialization (the appearance of ghosts at séances); astral bodies ("the immaterial part of the self that travels when the body remains still"); and reading of thought (mental telepathy).[117]

Psychical researchers believed they had demonstrated the existence of telepathy by the early 1880s. They were just as eager to prove the reality of the other phenomena mentioned above, and meticulously documented alleged ghost sightings and other occult happenings in the *Journal of the Society for Psychical Research.* The Society itself was founded in 1882 by a group of respected philosophers and scientists, some of them faculty members at Cambridge (including the Society's first president, Professor Henry Sidgwick). Other presidents of the Society in the first decades of its existence included such scientific luminaries as William James, who served as the SPR's first American president; anthropologist and folklorist Andrew Lang; and respected physicists Lodge and Barrett, both Fellows of the Royal Society.[118] The SPR's membership was equally illustrious, including British Prime Minister Arthur Balfour alongside scientists Pierre Janet, Cesare Lombroso, G. Stanley Hall, and Sigmund Freud.[119] Thus, one cannot easily dismiss the SPR as a loony fringe organization. In fact, during the late-Victorian and Edwardian periods, investigation into occult phenomena represented a semi-respectable branch of scientific research.

While many members of the SPR were respected scientists, they nonetheless reacted against certain materialist trends within mainstream scientific research, particularly the widening gap between science and spiritual concerns. The Society's stated goal, printed in its *Journal,* was "to examine without prejudice or prepossession and in a scientific spirit, those faculties of man, real or supposed, which appear to be inexplicable on any generally recognized hypothesis."[120] The impetus to found the SPR, Renée Haynes suggests, was likely "the conscious or unconscious acceptance of an analogy between living processes and the mechanical triumphs of the industrial revolution. This probably began with Descartes' contention that animals were automata."[121]

Like Stoker, members of the SPR shuddered at the suggestion that "consciousness was an epiphenomenon, thought was as it were the sweat of the brain," and worked hard to prove that the mind was more than mere mechanism.[122] They therefore encouraged scientific research affirming the existence of the soul. Though the group's members were predominantly agnostic rather than Christian, Janet Oppenheim points out their deeply felt desire to "find some other basis for the ethical precepts they cherished and some reassurance that all human suffering was not utterly devoid of purpose."[123] In other words, psychical researchers sought a surrogate belief system that would help reconcile tensions between religion and scientific materialism.

To cite one obvious example of these larger trends, founding SPR member Frederic Myers spent many years accumulating empirical evidence of the existence of an immortal soul. This evidence included the testimony of mediums who believed they had communicated with spirits of deceased individuals, in addition to studies of trance states and telepathic communications between living persons. Like Stoker, Myers was anxious for psychology to return to its roots as a science of the soul, so much so that he was often quite gullible in terms of the evidence he accepted. As Oppenheim explains, Myers desperately hoped that his findings would "win a secure niche for the human soul in the world of natural science."[124] Myers's cumulative work was published in the sprawling tome *Human Personality and Its Survival of Bodily Death* (1903). The perennial popularity of *Human Personality* testifies to the enduring need to find links between science and faith.

It is easy to imagine why Stoker, as a Christian trying to reconcile his beliefs with his scientific training, would have been attracted to the SPR, and to Myers's work in particular. Stoker attended at least one lecture given by Myers, and the two men were social acquaintances. For instance, Stoker reported attending "a delightful breakfast in the house of Frederick [*sic*] Myers" in Cambridge on June 16, 1898.[125] While I have not been able to ascertain whether Stoker was a member of the SPR, the author's interest in psychical research was well known and longstanding enough to garner some good-natured ridicule from *Punch*. A 1911 humorous piece entitled "How to Humanize the Landscape" connects Stoker with psychical researcher (and SPR president from 1901 to 1903) Sir Oliver Lodge:

A very touching act of homage has recently been paid to Mr. Bram Stoker. Simultaneously on the links at Stoke Poges and Bramshott two new pot bunkers, cut so as to represent the Olympian head (in profile) of the eminent novelist and impresario, have been dug in celebration of his fiftieth interview with Sir Oliver Lodge on the Psychical Significance of Vampires.[126]

If any such interviews took place, none were ever published, nor does Stoker mention Lodge in his most autobiographical work, *Personal Reminiscences of Henry Irving*. One must assume that *Punch* took creative liberties in order to highlight Stoker's well-known fascination with the occult.

Despite this good-natured ridicule, Stoker was hardly the only late-nineteenth-century writer to express such an interest. One thinks, for instance, of Sir Arthur Conan Doyle's turn to spiritualism or George Eliot's friendship with Myers. Stoker's friends, fellow novelists Samuel Clemens and Hall Caine, likewise shared his passion for the mysterious and inexplicable. Belford writes that Stoker and Clemens, who met in 1883 in Chicago, "exchanged views on the conflicts of duality, on nightmares, and on the unconscious."[127] When Clemens moved to London in the late 1890s, the two resumed their acquaintance and "their discussions of dreams and dual personalities continued," including conversations about *Jekyll and Hyde*.[128]

While Stoker's interest in the occult was not unusual in and of itself, it is surprising how few critics (Greenway excepted) have emphasized the importance of psychical research in *Dracula*. As Greenway explains, a comparison of the novel's two physicians, Van Helsing and Seward, reveals Stoker's views on the relative merits of normal science versus psychical research. While the two physicians hold similar degrees, Van Helsing outranks Seward professionally. Moreover, Van Helsing's occult theories prove to be correct more often than not, while Seward's scientific biases lead the vampire hunters astray. Stoker thereby signals his preference for a more open-minded approach to science than Seward's, one that combines philosophy, empathy, and self-control with mastery of the scientific method.[129]

Seward's commendation of Van Helsing's scientific virtues likely resembles Stoker's own opinion as to what constitutes "good" science:

> [Van Helsing] is a philosopher and a metaphysician, and one of the most advanced scientists of his day; and he has, I believe, an absolutely open mind. This, with an iron nerve, a temper of the ice-brook, an indomitable resolution, self-command and toleration exalted from virtues to blessings, and the kindliest and truest heart that beats … for his views are as wide as his all-embracing sympathies. (106)

As this passage suggests, Van Helsing embodies both the intellectual vigor of mainstream science and the values emphasized by the psychical researchers, including empathy and open-mindedness about extraordinary phenomena. He also exhibits masculine qualities like "self-command" alongside the Christian virtues of toleration and kindliness. If Van Helsing represents Stoker's ideal scientist, then Stoker apparently gravitated toward a more religiously conservative brand of science such as Carpenter's, and toward the occult inquiries practiced by the SPR.

To be sure, the SPR's investigations had much in common with mainstream science, and many illustrious researchers had one foot in each camp. Take, for instance, William James, who founded and served as president of the American branch of the SPR, while conducting some of the most respected psychological research of the late-Victorian era. Van Helsing seems modeled along the same lines. As Renfield explains, the Dutch physician has "revolutionized therapeutics by his discovery of the continuous evolution of brain matter" (215). (Apparently, Van Helsing's accomplishments are so widely known that even a hospitalized lunatic is aware of their significance.) Stoker's composition notes for *Dracula* reveal, however, that Stoker envisioned Van Helsing as a psychical researcher as well as a scientific expert. An early list of characters reveals that Van Helsing is an amalgam of three roles: a detective, a psychical research agent, and a German professor.[130]

Renfield's praise of Van Helsing's discoveries is one of the few moments in *Dracula* where we learn of the lead vampire hunter's medical specialty: neurology. This detail is significant for several reasons. Seward is also an expert in brains and brain disease, yet unlike Van Helsing, he professes enthusiasm for vivisection and even conducts experiments on patients, calling Renfield's case "a wonderfully interesting study" (110). When Renfield requests a cat to kill his collection of sparrows, Seward toys with the lunatic, stating, "I shall test him with his present craving and see how he will work out" (70). Later, Seward deliberately aggravates his patient by speculating about the soul of an elephant (238). On occasion, Seward feels remorse for baiting the lunatic: "In my manner of doing it there was, I now see, something of cruelty. I seemed to wish to keep him to the point of his madness – a thing which I avoid with the patients as I would the mouth of hell" (61). Seward's insensitive treatment of Renfield aligns the madhouse superintendent with controversial physiologists like Ferrier, Burdon-Sanderson, and Charcot.[131] Although Van Helsing operates on patients in grave emergencies (as with Renfield's brain surgery or Lucy's blood transfusions), he makes no mention of vivisection and seems to prefer observation of nature to interference with its processes. By making the kindly Van Helsing a brain expert, Stoker suggests that even physiologists can practice "good science" that neither interferes with religious belief nor harms helpless creatures.

Significantly, Victorian physiological psychology was a realm where mainstream science and physical research overlapped. This fact suggests yet another reason why Van Helsing should be an expert on the brain rather than, say, the heart or the pancreas. Late-Victorian psychical researchers investigated numerous phenomena of interest to psychologists, even if they reached different conclusions. For instance, both groups studied so-called automatic behaviors of the cerebrum (somnambulism, dreams, hypnotic trance states, dual personalities, etc.) using the scientific method. Mainstream scientists interpreted these behaviors as evidence of man's strictly earthbound nature, determining that people are mere machines. Huxley, for instance, compared Sergeant F.'s singing during his second state to the reflexive croaking of a frog whose fore-brain has been removed.[132] By contrast, psychical researchers felt that automatic mental phenomena displayed human potential to transcend the material universe. Myers, for instance, connected cerebral automatism to telepathy and clairvoyance. He viewed automatic mental processes, therefore, as an "introduction into a realm where the limitations of organic life can no longer be assumed to persist."[133]

Perhaps the greatest difference between the SPR and mainstream psychology, then, had to do with assumptions rather than methodology. While mainstream science nurtured a prevailing intellectual climate of skepticism, psychical researchers exhibited a resilient "will to believe" in incorporeal, spiritual phenomena.[134] Although both groups relied on the scientific method, their conclusions diverged because they started with such different hypotheses. Van Helsing typifies psychical researchers in his willingness to assume that material phenomena generally have spiritual significance. He tells Seward, the novel's skeptic, "I want you to believe" (172). Van Helsing views scientific skepticism as not merely misguided, but dangerous. Comparing the vampire hunters to "knights of the Cross" whose mission is to redeem souls, Van Helsing warns his associates that "in this enlightened age, when men believe not even what they see, the doubting of wise men would be [Dracula's] greatest strength" (278–9).

The novel itself is constructed in a manner that Van Helsing would surely endorse. Like the *Journal of the Society for Psychical Research*,

Stoker's *Dracula* consists of collected anecdotes about occult phenomena, written by allegedly "reliable" sources – mostly respectable middle-class men and women, plus the occasional nobleman or diplomat. *Dracula*'s multiple narrators span the same range of social classes. They also try their best to practice scientific objectivity, hoping that their emotions will not get in the way of accurately recording events. For instance, several journal entries by Seward, Jonathan, and Lucy begin with phrases such as, "Let me begin with facts," or "this is an exact record of what took place tonight," especially after anything supernatural has occurred (35, 130).

The novel's epistolary construction likewise conveys an impression of veracity. In addition to journal entries, newspaper clippings and telegrams interrupt the narrative on occasion, suggesting the reality and immediacy of the events described. Like a work of psychical research, the novel is an accumulation of evidence that might serve as "scientific" proof of the existence of phenomena such as immortal souls, free will, and even vampires. But whether *Dracula* succeeds in its lofty goals depends upon one's interpretation of the novel's concluding paragraphs.

## STAKING OUR VAMPIRE

On the surface, *Dracula*'s ending seems upbeat. When Dracula dies, it appears that Van Helsing's "good science" has rescued humanity from the evil physiology embodied in the bloodthirsty Count. The vampire hunters not only dispatch their enemy, but free his soul in the process. Mina relates that "even in that moment of final dissolution, there was in [Dracula's] face a look of peace, such as I never could have imagined might have rested there" (325). Moreover, the vampire hunters' full account of their quest is preserved for future generations, just in case these monsters rise to trouble mankind again. If one reads the novel from the perspective of late-Victorian physiological psychology and the discourse of the soul (as I have done so far), one might assume that our souls and free will are no longer threatened from without by the depredations of vampiric physiologists.

But the novel's pat conclusion crumbles under rigorous scrutiny. As Auerbach and Skal point out, Dracula's "final, hurried destruction is shorn of the rituals Van Helsing decrees necessary to drive vampires into the 'true death.'"[135] Dracula is dispatched with a knife, rather than a wooden stake, and his killers fail to decapitate him, leaving open the possibility that he will reincarnate. Furthermore, Jonathan poignantly realizes that the vampire hunters' masses of evidence may not really convince the world of the story's veracity:

> We were struck with the fact, that in all the mass of material of which the record is composed, there is hardly one authentic document; nothing but a mass of type-writing, except the later notebooks of Mina and Seward and myself, and Van Helsing's memorandum. We could hardly ask anyone, even did we wish to, to accept these as proofs of so wild a story. (326–7)

This mournful declaration must be read as an admission of failure. Even if the vampire hunters' supernatural adventures really happened, no one will believe them.

Jonathan's admission foreshadows the fate of psychical research in the late twentieth and twenty-first centuries. While interest in spiritualism and psychical research continued during and after the First World War, as bereaved families used mediums to try to contact deceased loved ones, the increasingly materialist emphasis of mid-twentieth-century science left little room for occult concerns. Today, psychical phenomena have become the exclusive domain of fiction and other popular entertainments, while scientists confine themselves to studying the material world. Meanwhile psychology, once a science of the soul, has fully transformed into a science of the brain. LeDoux writes, for instance, that scientists now see the brain and the self as one and the same: "the self is created and maintained by arrangements of synaptic connections."[136]

These conclusions seem as unnerving now as cerebral localization must have appeared to Victorians. Psychology, in its modern scientific form, vampirically deprives us of soul, will, and immortality. Try as we might, we have yet to stake our vampire, which may explain the need to compulsively repeat and relive his death in the countless vampire fictions written since 1897. Ultimately, *Dracula* and its descendants cannot entirely reassure us that we are more than the sum of our brain functions.

## CODA

Vampires are a remarkably versatile species whose permutations reflect changes in science and culture more generally, as Auerbach has persuasively argued.[137] How, then, do we make sense of the ongoing vampire craze, and its relationship to modern scientific developments? Recently,

Mark McGurl has noted that fiction – especially genre fiction about monsters such as zombies and vampires – continues to be an important means by which people grapple with the fallout of scientific determinism. Such fictions resonate with the bleak conclusions of neurological science: "The rise of the zombie in popular culture registers the same truth about human beings as does neuroscience … what should the novel do once consciousness has been physically 'explained?' What happens to the tradition of novelistic realism stemming from Austen when the reality is that we are all a bunch of tottering skin-bags animated by neural subroutines?"[138] Now, as in the late-Victorian period, novelists are turning to popular genres in order to address these provocative questions. The vampire, the zombie's glamorous relation, features prominently in such fiction, which readers eagerly devour.[139] Legions of fans await the next installment of popular vampire series like Stephenie Meyer's *Twilight* franchise and Charlaine Harris's Sookie Stackhouse novels (to name only the most recognizable examples of this genre). Vampires also populate contemporary film and television, notably in Alan Ball's *True Blood* series (an adaptation of Harris's novels) and in various film versions of the *Twilight* books.

Such fiction remains popular because vampires help us confront – and also evade – the realities of biological determinism. While modern scientists cannot promise us an existence beyond the grave, they can now offer viable methods of prolonging and enjoying our lives here on Earth, including means of slowing the aging process, enhancing our mood and appearance, and improving our sex lives. Much like plastic surgery or Viagra, the vampires envisioned by Harris or Meyer promise their victims eternal youth, immortality, and sexual potency. No wonder would-be victims often clamor to be bitten, despite lingering concerns over their souls or eventual salvation. For instance, Bella Swan, the protagonist of Meyer's *Twilight* franchise, begs her vampire boyfriend to make her eternally young like himself, even if it means sacrificing her chance at spiritual immortality. "I don't care, Edward … you can have my soul!" she cries, after a disturbing nightmare in which she imagines her aging body next to that of her perfectly preserved lover.[140] Unlike Dracula, who terrorized the British populace by threatening their souls, twenty-first-century vampires help us avoid facing mortality altogether.

All of this does not mean that people have stopped caring about souls, or ceased to worry about the pervasive effects of biological determinism on modern life. The lingering aura of terror surrounding vampires suggests that these philosophical problems have not magically resolved themselves, even if we no longer expect scientists to seriously engage with them. While the contemporary vampire craze may be as ephemeral as Marie Corelli's spiritualist blockbusters of the 1890s, we must still take this phenomenon seriously as an indication of the frighteningly close and ever-evolving relationship between vampires, science, and popular culture.

# PART II

# *Materialists*

CHAPTER 3

# *Photographic memory in the works of Grant Allen*

While Bram Stoker's spine-tingling vampire lore traded on fears inspired by cerebral automatism, some of Stoker's contemporaries enthusiastically embraced the idea of the human being as thinking machine. Grant Allen, an Anglo-Canadian novelist, journalist, and devotee of Herbert Spencer, opened his first published work with an elaborate comparison between the human body and a "locomotive engine." In *Physiological Aesthetics* (1877), Allen argued that steam engines differ from human beings in four principle ways: first, humans "have been evolved by natural selection, instead of having been consciously produced by the art of man"; second, humans are "self-conserving," since they can take care of their own needs for food, shelter, and other necessities; third, humans can reproduce, while locomotives cannot; fourth, humans are "endowed with consciousness."[1]

But whereas many Victorian writers would have emphasized this fourth and final point as the definitive distinction between man and machine, Allen saw consciousness itself as largely mechanical in nature. He viewed the mind as "a thinking machine … minutely constructed, inscrutable in all its cranks and wheels, composed of numberless cells and batteries, all connected together by microscopically tiny telegraph wires."[2] Allen likewise argued that pleasure and pain result from the correct or incorrect functioning of the bodily mechanism: in his view, pleasure stems from the "strong but normal excitation of fully-nurtured nervous structures," while consciousness of pain results either from "dismemberments of the body" or from "a general state of innutrition, either in the body as a whole or any of its component systems."[3] Allen further insisted that even subtle intellectual and aesthetic pleasures, such as enjoyment of literature and painting, could be attributed solely to the functioning of the nervous system, rather than to abstract entities like the mind, will, or soul. Allen's statements sound strikingly similar to Victorian supporters of vivisection who paraphrased René Descartes, arguing that "the cries of animals are but the working of the curiously-contrived machine, in which, when one portion is touched in a certain way, the wheels and springs concealed in the interior perform their work … there is no consciousness or feeling."[4] But Allen went further than Descartes, emphasizing that even *human* emotion and intellectual activity amounted to no more than the creaking of a rusty wheel.

Allen is surely an extreme example of a writer who embraced "human automatism" and all of its logical consequences. But he was far from alone. As we have seen in the previous chapter, prominent scientists like Thomas Henry Huxley, William Kingdon Clifford, and Shadworth Hodgson challenged Victorians to accept the notion that we are all automata.[5] Moreover, Allen's views on pleasure and pain derived in part from the work of Alexander Bain, an influential psychologist and founder of the journal *Mind: A Quarterly Review of Psychology and Philosophy* (1876–present), who had previously argued that "states of pleasure are concomitant with an increase, and states of pain with an abatement … of the vital functions."[6] In fact, Allen's mechanical, brain-based view of human consciousness represented the extreme end of an ideological spectrum, at the other end of which were William Carpenter, Oliver Wendell Holmes, Bram Stoker, and others who refused to reduce mind, will, or soul to brain.

Allen's extreme physiological reductionism carried over into his novels and short stories of the 1880s and 1890s. Allen's biographer, Peter Morton, relates that Allen took to fiction reluctantly in order to supplement the scanty earnings brought in by his scientific writing.[7] Yet Allen's female detective fictions of the 1890s rank among his most appealing and thought-provoking writings. Novels like *What's Bred in the Bone* (1891), *Recalled to Life* (1891), *Miss Cayley's Adventures* (1899), and *Hilda Wade* (1900) blended aspects of New Woman writing with the elements of the detective story as perfected by Allen's close friend, Sir Arthur Conan Doyle. Critics have largely ignored these works in favor of Allen's scurrilous exposé of sexual mores in *The Woman Who Did* (1895).[8] Yet outside of Allen's scientific writings, these detective novellas are where Allen's ideas about the physiology of vision find their fullest expression.

Recent critics have emphasized that detective fiction is an expressly visual medium whose unprecedented popularity during the 1880s and 1890s paralleled the concurrent rise of cinematic technologies. Ronald Thomas describes how Sherlock Holmes and other fictional detectives solve cases using visual methods derived from late-Victorian medicine and criminology. For Thomas, both the detective and the physician are "master diagnosticians" capable of reading pathological symptoms overlooked by others: "Invariably, the mangled corpse the literary detective scrutinizes reveals a code that his trained eye is uniquely capable of reading; or, alternatively, the body of the suspect betrays its own guilt in some visible signs that are only legible to the eyes of the detective."[9] Lawrence Frank, meanwhile, finds parallels between detection and other sciences, namely geology and biology, disciplines that likewise require acute visual perception: "the geologist and the detective … inhabit a world of signs that to the unseeing eyes of others do not exist."[10] Finally, Elizabeth Carolyn Miller reminds us that Conan Doyle's detective stories (like many of Allen's) typically appeared in illustrated venues like George Newnes's *Strand Magazine* (1891–1950).[11] The drawings accompanying detective stories in these venues encouraged readers to test their own visual acumen against that of fictional detectives.[12]

It makes sense that such a visually oriented medium appealed to writers with extensive knowledge of the physiology of vision. Conan Doyle was a physician with advanced training in ophthalmology, while Allen explored the structure of the eye in scientific works like *Physiological Aesthetics* (1877) and *The Colour Sense* (1879).[13] Accordingly, their detective fictions deploy what Miller has described as a "relentlessly visual logic" that "echoes timely scientific procedures and principles."[14] As Sherlock Holmes explains in *A Study in Scarlet* (1887), a good detective can identify a man's calling by his "finger nails, by his coat-sleeve, by his boots, by his trouser knees, by the callosities of his forefinger and thumb" or other minute visual cues.[15] Similarly, Una Callingham, the amateur detective protagonist of Allen's *Recalled to Life*, identifies her father's supposed murderer solely based on the shape of his hands.

Una and Holmes share more than their surprising visual acumen. Both Conan Doyle and Allen enthusiastically compare their protagonists to machines, particularly cameras. At the beginning of "A Scandal in Bohemia," published in the same year as *Recalled to Life*, Watson describes Holmes as "the most perfect reasoning and observing machine that the world has seen," calling him a "sensitive instrument" with "high-power lenses."[16] Holmes thus appears to be "a kind of camera," as Thomas suggests, although one might also justly compare him to a microscope or other mechanical visual aid.[17]

While Conan Doyle occasionally compares Holmes to a machine or a camera, these metaphors are far more explicit and pervasive in *Recalled to Life*, which likewise includes a more detailed exploration of the physiology of vision than any Sherlock Holmes tale. In this novella, Allen depicts his heroine as a human camera whose eye takes pictures of scenes that register as photographic "negatives" in her brain. Una's precise memory of these images renders her brain akin to a photo album or gallery. Her visual registration of her surroundings relates to Allen's scientific views about the primacy of vision over other sensory modes. Moreover, the similarities between Una's photographic "mechanism" and the latest photographic technologies available in 1891 once again suggests Allen's persistence in equating bodies and machines.

Equating the eye with a camera and the brain with a gallery was a potentially radical maneuver, but one for which there were important scientific precedents. In Romantic and Victorian evolutionary debates, various writers described the eye as a "perfect" structure too complex to have evolved by means of evolution alone, yet still analogous with mechanical visual aids such as the telescope. In 1802, William Paley argued in *Natural Theology* that complex and beautiful natural structures must necessarily be produced by a divine Creator. By way of analogy, Paley suggested that "there is precisely the same proof that the eye was made for vision, as there is that the telescope was made for assisting it."[18]

Charles Darwin, who had read Paley's treatise at Cambridge, pointedly alluded to this passage in *The Origin of Species* (1859), where "a perfect and complex eye" serves as an unlikely example of natural selection at work:

> To suppose that the eye, with all its inimitable contrivances for adjusting the focus to different distances, for admitting different amounts of light, and for the correction of spherical and chromatic aberration, could have been formed by natural selection, seems, I freely confess, absurd in the highest possible degree. Yet reason tells me … that a perfect and complex eye could be formed by natural selection.[19]

Through a series of examples, Darwin goes on to show that the simple eyes of mollusks, consisting of "an optic nerve merely coated with pigment," could have evolved by numerous gradations into "a structure even as perfect as the eye of an eagle."[20] Darwin again evokes Paley's treatise when he adds: "It is scarcely possible to avoid comparing the eye to a telescope." Yet Darwin concludes that a "living optical instrument" must be "as superior to one of glass, as the works of the Creator are to those of man."[21]

On both sides of nineteenth-century evolutionary debates, then, the eye served as a lightning rod for controversy and a supreme example of either God's benevolent design or natural selection at work. Traditionally, vision has been viewed as an exalted faculty, with the eyes and optic nerves mediating contact between the soul and the outside world.[22] To invoke the eye in the context of a biological argument, therefore, was a sure way to attract attention and raise hackles, a fact that Darwin and Paley enthusiastically exploited for their own ends.

Allen's extended focus on vision in *Recalled to Life* is a similar maneuver, cleverly designed to draw attention to his pet theories about the mechanical nature of human consciousness. Allen, an enthusiastic proponent of both Darwinian and Spencerian evolutionary theories, went further than Darwin himself by suggesting that the "perfect" eye was entirely analogous with man-made mechanical structures. Allen's physiological reductionism stemmed partly from his admiration of Huxley, who held similar beliefs about the human brain as mechanism.

But more personal factors were also involved. Unlike Conan Doyle, whose scientific rationalism was tempered by his belief in spirits and fairies, Allen mistrusted supernatural explanations of any sort.[23] Allen viewed Christianity as a pernicious myth, believing that everything in the universe could be explained by scientific laws alone. Morton relates how Allen's sole work of anthropology, *The Evolution of the Idea of God* (1897), aimed to expose the "folk-myth origins" of Christianity and thereby "help consign Christianity and all other revealed religions to the scrap heap."[24] Concepts like the soul or will doubtless struck Allen as musty relics of an outdated belief system. Accordingly, in *Recalled to Life*, Allen enthusiastically debunks the popular idea that eyes might serve as "windows to the soul," suggesting instead their purely mechanical function.

While Allen's atheism ran counter to mainstream Victorian thinking, his rhetoric at times curiously resembles that of his most vocal Christian opponents. Allen's acquaintances and biographers suggest that he was as zealous and unwavering in his atheism as any ardent religious believer. Astronomer and journalist Richard Proctor may have been thinking of his friend Allen when he wrote that "dogmatic denials" of God's existence could sound strikingly similar to "dogmatic assertions of belief."[25] Morton, meanwhile, describes Allen's quasi-evangelical desire to convert others to his own "religion," namely, "scientific naturalism."[26] As a result, Allen's mechanistic metaphors of human cognitive functions can sound curiously triumphant, as though he delighted in unmooring his readers' religious faith. In light of these facts, Allen's equation of the eye with a camera in *Recalled to Life* cannot be innocent or accidental. By depicting Una as a kind of superior machine, Allen was throwing down the gauntlet, challenging his readers to accept his own firm belief that human consciousness was entirely mechanical in nature.

Unsurprisingly, Allen's *Recalled to Life* is rife with biomechanical analogies. Since this novella is so rarely read or discussed by critics, a brief summary seems in order here. Narrator Una Callingham suffers from amnesia and temporary aphasia (loss of verbal ability) after witnessing her father's violent murder. She suddenly enters into a "second state" or "second babyhood" in which she remembers nothing of her previous life, not even her native tongue.[27] This state is likened to dual personality when journalists refer to Una as the "two-souled lady" (70). After recovering the use of language, Una attempts to reconstruct her forgotten life history by examining photographs from her past. She possesses a comprehensive pictorial record from which to piece together her former existence, left behind by her photographer father, Vivian Callingham. Una explains, "My father had been interested in chemistry and photography, it seemed, and had lately completed a new invention, the acmegraph, for taking successive photographs at measured intervals of so many seconds by electric light" (19). These photographs represent humans and animals in sequential stages of rapid movement, typically "men running and horses trotting" during athletic contests (50).

By collecting and sequencing her father's photographs, Una does much more than reconstruct her pre-existing memories. She also discovers that her father had concealed facts about her early life and true parentage. After stranding Una's real father on a desert island, Vivian Callingham married Una's wealthy mother, who believed her first husband to be deceased. Callingham's misdeeds lead indirectly to Una's own criminal behavior. In the course of her investigations, Una learns that she herself murdered her father in order to stop him from killing her fiancé, Dr. Jack Ivor, who had accidentally stumbled upon her father's secret criminal past. In the novella's dramatic conclusion, Una tracks down Dr. Ivor, who has gone into hiding in rural Canada. He helps Una piece together the crucial elements of the mystery that have so far eluded her, including the true identity of the murderer (Una) and her motive for killing her father. Though Una briefly considers turning herself in to the police, the lovers ultimately decide to marry and to conceal their shared knowledge of Una's crime.

In this chapter, I will explain how and why Allen's attempt to equate his heroine with a mechanism – specifically, a camera – only partially succeeds. Unlike Sherlock Holmes, a machine-like character who rarely makes a mistake or betrays an emotion, Una's first-person narrative presents her terrifying subjective experiences in ways that emphasize her vulnerability and suffering. Due to the novel's subtle equation between photographic registration and traumatic shock, the reader is left with a haunting memory of Una's feelings of loss, rather than an objective view of her mental "machinery" at work. No matter how relentlessly Allen compares Una to a photographic apparatus, one cannot help feeling that her intense psychological pain is *not* equivalent to the creaking of a rusty wheel.

Paradoxically, the very techniques Allen uses to portray his heroine as a mechanism often add psychological depth to her character. In the course of the narrative, Una receives numerous psychological shocks, after which she invariably feels faint, trembles like an aspen leaf, turns pale, or shrieks in horror. Here are just a few examples of such episodes, which occur on nearly every page:

"With a horrible revulsion I realized the truth." (87)

"At that moment of supreme horror, a thrill ran all through my body." (90)

"With an awful recoil, I drew back and suppressed a scream." (63)

"In a moment a terrible sinking came over me from head to foot. I trembled like an aspen leaf." (47)

These somatic responses to psychologically distressing discoveries have a machine-like, action–reaction sort of character, like the physiological responses to sensory stimuli described in *Physiological Aesthetics*. Una herself describes such reactions as "instinctive" or automatic (51). Yet these repetitive episodes also underscore the very severity of Una's psychological distress, not to mention her helpless vulnerability during her amnesiac state. Arguably, these episodes add dimension to Una's character rather than flattening it out, because they help us empathize with her painful quest to recover her lost identity.

Moreover, the dreamy strangeness of Una's recovered memories contributes an aura of the supernatural that undermines the novella's scientific rationalism. After dreaming about her forgotten childhood in Australia, Una "sat up in bed, appalled … if I had seen a ghost, it could hardly have affected me more profoundly than this ghost of my own dead life thus brought back suddenly to me" (43). Similarly, when Una attempts to recall the crime scene, she sees a "phantom hand … that floated like a vision, all distinct before my mind's eye" (47). As in Conan Doyle's Sherlock Holmes stories, all such apparently supernatural instances turn out to have a rational explanation once the mystery is solved. But the heroine's feelings of being "pursued and haunted" by her visions, and her repeated assertions that all her experiences since the murder are "like a dream," undermine the stable sense of reality upon which scientific thinking – and detective stories – are supposed to rest (44, 30).

Allen's connection between trauma and mental automatism drew heavily upon late-Victorian research on mental illness, particularly Théodule Ribot's *Diseases of Memory* (1881), which he read and admired.[28] Allen was also scientifically up to date in linking dual personality and what we would now call dissociation – which can involve feeling disconnected from reality, as if one is living inside a dream or seeing the world through a veil.[29] But ironically, Allen's commitment to psychological accuracy is precisely what makes Una seem human rather than machine-like. Indeed, Una's descriptions of her experiences might sound eerily familiar to anyone who has ever endured a traumatic shock, felt somehow detached from reality, or remembered forgotten events while dreaming.

For all its scientific accuracy, then, Allen's novel inadvertently slides into the realm of Gothic mystery. This fact testifies to the slippage between the related genres of detective fiction, sensation novels, and the Gothic, not to mention the inherently Gothic qualities of vision and its mechanical equivalent, photography. While both the eye and the camera are commonly associated with objectivity and scientific accuracy, both structures came under scrutiny in the nineteenth century due to their obvious *imperfections* and the ways in which these flaws distorted human experience. Allen's chosen genre and his biomechanical metaphors would seem to place him on solid scientific ground, yet these are exactly the elements that allow his fiction to run away with him. Thankfully, *Recalled to Life* gains depth and richness due to this apparent failure of scientific rationalism to dispel the aura of mystery surrounding photography, vision, and human consciousness.

## THE HUMAN CAMERA

In the novella's opening pages, Una explicitly compares her retina to the camera's "sensitive-plate," underlining the parallels between human biological structures and photographic apparatus (5). At the moment of the trauma, Una's optic nerves take a picture of the "Horror" she has witnessed: "No sensitive-plate could have photographed [the scene] more instantaneously, as by an electric spark, than my retina that evening, as for months after I saw it all" (5). She explains that "the Picture" left behind by the trauma "has burned itself into my brain like a flash of lightning!"

just as swiftly and surely as her father's flash photography captures images on sensitive plates (21).

This analogy between photography and sensory perception was not unique to Grant Allen's fiction. Photographic metaphors were in fact relatively common in nineteenth-century scientific writing and, to a more limited extent, in imaginative fiction. Douwe Draaisma explains: "From the mid-nineteenth-century onwards, when developments in photochemistry and improvements in the construction of cameras produced sharp photographs, photographic metaphors of all kinds appeared in papers on the visual memory, gradually changing the human brain into a light-sensitive plate, the memory into an album full of silent snapshots, consciousness into a gallery." Draaisma emphasizes that for Victorians, photographs and phonograph sound recordings served as "artificial memories" that in turn "shaped … views of remembering and forgetting."[30] Laura Otis has likewise emphasized how biomechanical metaphors involving sensory perception permeated Victorian consciousness. Comparisons between mechanical and organic communication systems were common throughout the nineteenth century. In fact, the inventors of the telegraph and telephone and physiologists studying the nervous system looked to each other for inspiration.[31] For example, Alexander Graham Bell studied a human ear procured by a physician friend in order to grasp the mechanics of sound transmission and perception.[32]

While the use of photographic metaphors for visual processes was common in scientific writing, in *Recalled to Life*, the connection between physiology and photography goes beyond metaphor. Una doesn't need a camera to take pictures of the crime scene, because she *is* a camera whose retina registers photographs just as efficiently as her father's apparatus. In one memorable scene, the accuracy of Una's mental "photography" is tested by a Scotland Yard officer. Una can describe the crime scene to the officer precisely, consulting only her internal, mental picture of "the Horror" (14). Unbeknownst to Una, the officer holds a photograph of the crime scene, which he then shows her. She registers her "awe and amazement" that "it was indeed the very scene I remembered so well … everything in the room was, allowing for the changed point of view, exactly as I remembered it in my persistent mental photograph" (16).

The idea that Una's mental "picture" might serve as legitimate forensic evidence calls to mind the popular but fallacious idea of the retinal after-image as an aid to solving crimes. From the 1860s until well into the twentieth century, a pervasive folk belief suggested that photographing the eyes of a murder victim could reveal the identity of the murderer. This theory of the "tell-tale eye," as Véronique Campion-Vincent memorably terms it, had some scientific basis in a series of late-Victorian optical studies. In the late 1870s, physiologists F. Wilhelm Kühne and Franz Böll discovered that the retinas of recently deceased animals, when stored in a solution of alum, could retain images seen shortly before or after death. The first such experiment was conducted by Böll, who famously discovered the "retinal violet" (now known as rhodopsin) in 1876.[33] Böll proposed that this light-sensitive, pigmented retinal membrane registered visual images through photochemical processes. This discovery was hailed as "a veritable revelation" by later scientists, who viewed the liminal membrane as the crucial point of interaction between the brain and outside world.[34] Some even explicitly compared the retinal purple to "the sensitive plate in photo cameras," highlighting the blurry boundary between sense organs and visual technologies.[35] Allen clearly knew of the existence and functions of the retinal purple, to which he referred in *The Colour Sense* and elsewhere.[36]

While Böll's experiments revolutionized ophthalmology, his findings were sometimes misapplied by late-Victorian forensic experts. Following the infamous Jack the Ripper murders in the late 1880s, for instance, the eyes of several victims were removed and photographed in the hopes of revealing the murderer's identity, but without success.[37] As early as the 1860s, educated readers of newspapers debunked the "tell-tale eye" theory by pointing out that visual impressions remain on the retina for only fractions of a second, and that the eye decays rapidly after death.[38] But despite these objections, the theory continually resurfaced in late-Victorian crime journalism, and continues to appear in crime fiction from time to time.[39] Its enduring popularity may perhaps be traced to another folk belief, the idea that the eye is the window to the soul.[40]

Ever the contrarian, Allen loved debunking popular mythologies of this type, as he demonstrated in stories like "A Mysterious Occurrence in Piccadilly" (1884) and "Our Scientific Observations of a Ghost" (1878), both of which mock the widespread *fin-de-siècle* fascination with psychical research.[41] In a similar vein, *Recalled to Life* subtly ridicules the popular notion that the retinal after-image could be used as reliable forensic evidence. The Scotland Yard officer who questions Una proceeds under false assumptions, believing that he is interviewing an innocent bystander who has seen the murderer get away. In reality, he is interviewing the murderer, who saw the fleeting image of an innocent bystander (Dr. Ivor) escaping from the window. Thus, while Allen did not appear to question the similarities between the retina and the sensitive plate, he apparently objected to the misuse of Böll's discovery within forensic science.

While the retinal purple was seen as directly analogous to the photographer's sensitive plate, *Recalled to Life* also suggests more subtle parallels between photographic memory and recent photographic technologies. Specifically, the automatic nature of Vivian Callingham's acmegraphic process resonates with theories of cerebral automatism in vogue at the time, which posited that "unconscious cerebration" or "reflex action of the cerebrum" accounted for many types of mental associations and thought processes. Just as Una's father's "automatic machine, the acmegraph" had "produced all unconsciously the picture of the murder," her brain "photographs" the crime scene effortlessly yet effectively (50).

Many of Una's mental functions have this automatic character. She relates how her "intellect, working unconsciously and spontaneously in an automatic condition" conjures up mental snapshots of her past (42–3). This makes sense, as late-Victorian physicians associated amnesia with various forms of cerebral automatism. Ribot argued in *Diseases of Memory*, for instance, that "the most highly developed and most unstable forms of memory" were the first to be lost in amnesiac conditions, often leaving the automatic behaviors (eating, speaking one's mother tongue, tying one's shoes) largely intact.[42] Anything learned in early childhood was more likely to be retained, Ribot suggested, as was any lesson or behavior repeated so often that it became automatic rather than conscious.

By creatively combining Victorian biology and photography, *Recalled to Life* demonstrates the ease with which even the most complex mental processes could be reductively described via mechanical metaphors, especially when the author was untroubled by the moral or theological implications of such descriptions. Biomechanical metaphors involving the eye as camera and the brain as light-sensitive plate seemingly underscore the widely held notion that photographs present a faithful reproduction of truth or an acceptable substitute for human memory. But Allen's novella complicates the easy equation between photography and truth by suggesting that any given photo can only present a partial or fragmentary reality.

The whole truth – and the solution of Una's mystery – only emerges when she arranges her photographs to present a composite image of her pre-amnesiac existence. For instance, Una discovers the identity of her fiancé by juxtaposing photographs of hands, shoulders, and other eerily disconnected body parts appearing separately in various photographs. Taken together, her assembled photographic memories produce a collage effect reminiscent of mid-Victorian "composition photography" and "art photography," in which numerous photographic negatives were pasted together to produce sublime visual compositions.[43] These composition photographs aimed not at truthful representation, but at artistic effects similar to those achieved by Victorian painters. Like the fragmented bodies that appeared in such photographic artworks, Una's photos present partial versions of the truth that she must combine and cognize in order to create a fuller record of her fractured history. The process by which she integrates her psychologically distressing, disjointed memories will receive greater attention toward the conclusion of this chapter.

## THE EVOLUTION OF PHOTOGRAPHIC MEMORY

So far, I have suggested how Allen's view of the retina as sensitive plate drew upon (and occasionally contradicted) the received scientific wisdom of his contemporaries. Here, Allen's own scientific writings on vision will receive extended attention, particularly those works in which he extols vision as the principle sensory mode guiding human behavior. Allen's strong opinions about the supremacy of vision help to explain why *Recalled to Life* features a human camera, rather than, say, a human phonograph or telegraphic switchboard – both of which devices served as alternative metaphors of memory in the late nineteenth century.[44]

Allen was intimately familiar with Victorian scientific literature on physiological psychology and sensory perception. Moreover, he was justly respected as a synthesizer and popularizer of physiological theories, and to a more limited extent, as an innovator in his own right.[45] Yet, as Suzy Anger has noted, critics have largely overlooked Allen's interest in psychology and physiology, paying far more attention to his controversial views on evolutionary biology and women's rights.[46] This oversight is puzzling given that Allen began his career as self-proclaimed psychologist.[47] His first publication, *Physiological Aesthetics*, was an ambitious but ultimately unsuccessful attempt to outline the neurological basis of artistic sensations. Although no author, then or now, has truly fathomed this mysterious subject, Allen's erudite attempt received positive reviews. Allen's second major work, *The Colour Sense*, established his reputation as an expert on the evolution of vision. These early works, combined with a series of articles Allen wrote for the journal *Mind* in the 1870s and 1880s, earned him the respect and admiration of intellectual luminaries including Darwin, Huxley, and Allen's hero, Herbert Spencer, whom he described as "the maximum brain on earth."[48]

Of these early publications, Allen's articles in *Mind* reveal the most about his abiding interest in psychology, retinal structure and function, and the innervation of vision in the brain. Allen's 1881 article, "Sight and Smell in Vertebrates," shows the author's support for cerebral localization. In this article, Allen cited the work of Scottish neurologist David Ferrier, whose experiments on the brains of live monkeys proved that electrical stimulation of certain cerebral areas predictably caused specific physical reactions. Ferrier's experiments helped overturn the older view that the cerebrum consisted of a reticulum or nerve network that functioned holistically to produce thoughts, emotions, and muscular responses. By aligning himself with localizationists, Allen demonstrated his progressive thinking on brain function. He acknowledged that sight, like other mental faculties, was localized in a particular region of the brain rather than spread throughout the cerebral cortex.

This article and others like it demonstrate Allen's view that vision is the most important and highly evolved sensory mode in human beings, and that improvements in vision go hand in hand with increased intelligence. This opinion was not shared by all Victorian scientists. Spencer, for instance, wrote that "touch, the simplest and earliest sense, should, in its higher forms, be more than any other sense associated with the advance of intelligence," although he saw vision and touch as closely linked faculties.[49] By way of example, Spencer noted that intelligent animals (especially parrots and elephants) are skilled at grasping food and other objects, and that in humans, "highly developed manual dexterity" is needed to conduct scientific experiments.[50] Darwin, meanwhile, saw both vision and hearing as superior to the olfactory sense, which he thought existed "in an enfeebled and … rudimentary condition" in human beings, especially civilized Europeans.[51] Maudsley and Ribot, like Allen, both saw vision as the sensory faculty that civilized Europeans relied on the most.

Even more than these prominent thinkers, Allen emphasized the degree to which vision was a highly evolved, intellectual faculty that makes us truly human. Throughout the animal kingdom, Allen wrote, "advance in visual adaptation" generally goes along with "advance in intelligence."[52] By contrast, Allen argued that the sense of smell predominates in lesser beings, including savages, "idiots," and lower animals, all of whom "use olfactory sensations as practical guides to an extent quite unknown amongst the higher races."[53] "As we mount the vertebrate scale," Allen argued, "we find sight gaining in relative importance, and smell losing in relative importance."[54] By way of example, he cited dogs' excessive reliance on smell, not to mention "idiots … [who] are in the habit of smelling at food and other objects given to them."[55]

In contrast to those who viewed language (and thus hearing) as the most highly evolved human faculty, Allen countered that "visible symbols are the language of our thought … our world is a picture."[56] By describing human memories as a series of pictures, Allen foreshadowed the plight of his protagonist Una Callingham, who must reconstruct her forgotten memories entirely from photographs. Although recovering language is the first step in Una's recovery, she must then assemble a *visual* vocabulary in order to make sense of her environment. Her mind is literally a picture book.

Despite Una's diseased mental state, Allen implied in his scientific writings that all human brains are likewise fashioned to retain and sequence visual data. He provided the example of a man who must find his way from "Hyde Park Corner to Oxford Circus by means of a whole string of objects, observed and recollected as signs of the road."[57] In this instance, as elsewhere in Allen's work, sequenced visual data – rather than auditory or olfactory cues – provide the most important guides for human conduct.

Allen cited physiological data that supported his opinions about the relative importance of different sensory faculties. He argued that we rely more on vision than our senses of hearing, touch, or taste, because vision is the most innervated sense in the human brain. "If we cut open the head of a man," Allen wrote, "we shall find in it a large and highly developed optic centre, directly connected with the eye and the nerves of sight, and having numerous side connections with other parts of the brain. This large nervous mass accurately reflects the extreme importance of sight in the human system." On the other hand, Allen stated, "if we cut open the head of a dog, we find … a very big and very important olfactory lobe, having an immense number of lateral connections with every other part of the brain."[58] By engaging in this imaginary vivisection of a human being and a dog, respectively, Allen aligned himself not just with the theories but also with the controversial methods of cerebral localizationists, who removed or electrified portions of live animals' brains to observe the resulting behavioral dysfunctions.[59] Such practices were publicly exposed during Ferrier's well publicized 1881 trial, in which he was accused but ultimately acquitted of violating the 1876 Anti-Vivisection Act.

Allen's point of view may have been influenced by his own extraordinary visual memory. Morton remarks upon this astonishing faculty, noting that Allen (who was also an excellent botanist) "could identify forty thousand plants by eye alone."[60] Similarly, Allen's friend Richard le Gallienne recalled "the prodigious range and accuracy and instantaneous readiness of [Allen's] memory. This was so proverbial amongst his friends that one of the dearest of them coined the phrase, 'We must look it up in Grant.'" According to le Gallienne, Allen once joked about "abandoning literature and setting up as a peripatetic encyclopedia … a sort of general call-office of knowledge."[61] Le Gallienne emphasized the visual nature of Allen's memory, suggested by the author's favorite "optical toy," an expensive "pocket microscope" given to him as a gift by Darwin and other illustrious members of the scientific community. Le Gallienne recalled that Allen "used constantly to twirl and twirl [this microscope] between his finger and thumb as he talked … without which I really think he could not have talked at all."[62]

Allen's prized microscope might serve as a convenient symbol of the minute, precise, and expressly visual character of his memory. Allen shared this trait with Sherlock Holmes, suggesting the possibility that Conan Doyle drew inspiration for his famous detective from his eagle-eyed friend. In any case, both Holmes and Allen apparently possessed what is controversially known as photographic or eidetic memory, a phenomenon commonly seen in children and autistic savants that only rarely manifests in adults.[63] Una Callingham's pictorial memory reconstruction provides us with an unusually literal form of this rare condition of extremely accurate visual recall.

While the term "eidetic memory" was introduced only in 1922, the scientific study of visual memory savants began in the 1860s and intensified during the decade when Allen wrote *Recalled to Life*. Intriguingly, mid- to late-Victorian discussions of photographic memory tended to link this condition to mental illness instead of extraordinary mental acuity. In his 1878 volume *Physiology and Pathology of Mind*, Henry Maudsley described a patient in the Earlswood Asylum for Idiots who could "repeat accurately a page or more of any book which he has read years before, even though it was a book he did not understand in the least." Maudsley compared this man's impressive visual memory to "a photographic copy of former impressions with his mind's eye." Such a "photographic" memory, Maudsley claimed, was "not indeed commonly associated with great intellectual power," but instead surfaced most frequently among idiots and the insane.[64] Three years later, Ribot described "certain idiots, unable to make the most elementary arithmetical calculations" who could nonetheless "repeat the whole of the multiplication tables without an error."[65] Yet Ribot also mentioned mnemonically gifted individuals who were not mentally impaired, including "artists like Horace Vernet and Gustave Doré painting a portrait from memory … chess players able to carry on one or several games in the mind … [and] lightning calculators … who 'see the figures before their eyes.'"[66] Allen, who read and admired Maudsley as well as Ribot, would surely have known of such examples.[67]

In the last decade of the nineteenth century, researchers investigating photographic memory shifted their focus from the mentally impaired to the intellectually gifted. This makes sense when one considers that many nineteenth-century writers, including Francis Galton, John Ferguson Nisbet, and Jacques Moreau, felt that "the idiot … the madman, as well as the genius, are … branches growing from the same tree."[68] In 1894, French psychologist Alfred Binet conducted a study of mathematical prodigies and blind chess players. Binet, who is now best known for his groundbreaking work on intelligence testing, focused on the visual memory strategies contributing to the success of these savants.[69]

Not surprisingly for a novella written between Maudsley's and Binet's studies, *Recalled to Life* suggests a close connection between photographic memory, mental illness, and extreme intelligence. Una is not only a bright, capable detective, but also a "Psychological Phenomenon," a "two-souled lady" whose divided consciousness could be classified as a type of insanity (9, 70).[70] Indeed, one might justly compare Una to cases of dual personality discussed in the late-Victorian British press, including Félida X.[71] But doctors at medical conferences instead describe Una as a cross "between Constance Kent and Laura Bridgman," an alleged murderess and a deaf-blind child prodigy, respectively (86). The latter comparison calls to mind the blind chess players Binet would later investigate, not to mention Ribot's view that aphasiac patients resembled deaf-mutes in certain respects.[72] Una's photographic memory is thus part of a larger pathological profile that includes not only dual consciousness, but also aphasia, amnesia, and possible moral insanity.

A few years later, Allen placed less emphasis on the connections between photographic memory and mental illness, following a broader trend among *fin-de-siècle* psychologists. For instance, his last female detective novel, *Hilda Wade* (1900), features a wholesome, down-to-earth female detective.[73] Toward the beginning of the novel, Hilda briefly suggests a connection between photographic memory and pathology when she remarks, "Unfortunately, I *can't* forget. That is a sort of disease with me."[74] Her statement mirrors Ribot's claim that "forgetfulness, except in certain cases, is not a disease of memory, but a condition of health and life."[75] Yet unlike Una, Hilda exhibits no other signs of mental instability throughout the novel. The narrator describes her as "a bright, well-educated, sensible, winsome, lawn-tennis-playing English girl" whose main peculiarity is her zeal for amateur detection.[76] Incidentally, she is also an accomplished amateur photographer, while her future husband is a physician who dabbles in proto-cinematic technologies.[77]

Yet for Allen, as for Binet, men of science (and not their female counterparts) served as the true benchmarks of superior mnemonic aptitude. This gender bias is evident throughout Allen's detective fiction. For instance, Hilda inherits her impressive memory from her father, a physician with legendary diagnostic powers, while Una's mental gifts may owe something to the influence of her stepfather, a chemist and photographic pioneer.[78] Allen considered men of science to be the most visually oriented human beings, and thus the most evolutionarily advanced. In "Sight and Smell in Vertebrates," Allen wrote:

> If we contrast the wild carnivore, sniffing the track of its prey, scenting its mate and young, taking stock of all nature by smell ... with the man of science, accurately measuring everything in the last resort by an appeal to the delicately discriminative sense of sight, employing microscopes and micrometers, dividing thermometers into visibly graduated decimals of degrees, testing chemicals by means of visible reactions, reducing the vague indications of all other senses – touch, taste, smell, heat and cold, pressure, muscular tension – to the definite indications which sight alone can yield; we shall see how absolutely essential is the predominance of sight for the evolution of the highest intelligence.[79]

By thus linking sight, evolutionary fitness, and scientific pursuits, Allen set up a recipe for eugenic perfection that resurfaces in *Recalled to Life* and *Hilda Wade*. Both works conclude by pairing off visually oriented women with men of science, physicians Dr. Ivor and Dr. Cumberledge, respectively. In Allen's view, this was exactly the sort of pairing likely to result in the most highly evolved human offspring.

Because the above passage emphasizes the use of tools such as Allen's beloved microscope, it subtly suggests that intelligence has a tactile as well as visual component. While Allen argued that sight is "the real privilege of man," he was also willing to acknowledge the secondary importance of touch, theorizing that "the hand and eye, going together, have made man what he is."[80] Possibly this was a concession to his intellectual hero, Spencer, who believed in the primacy of touch and its connection to evolutionary fitness. Allen elaborated, "If you and I see any object which we do not know, and if we are anxious to learn more about its nature, we go up to it and handle it," suggesting a strong association between grasping an object and fully cognizing it.[81]

In this respect, Allen's evolutionary philosophy resembles that found in H. G. Wells's 1893 article "The Man of the Year Million," in which Wells predicted that future humans will have "enormous brains, liquid, soulful eyes, and large hands," because the relative importance of intelligence, sight, and manual dexterity will have rendered other bodily functions unnecessary. The highly evolved Martians in Wells's 1898 novel *The War of the Worlds* resemble future humans in that they have numerous tentacles (in place of hands) and resemble "heads – merely heads. Entrails they had none." A dissection of the Martians reveals that their brains sent "enormous nerves to the eyes, ear, and tactile tentacles," showing that both vision and touch are highly innervated in the Martian cerebral cortex, with other senses being relatively unimportant.[82] In fact, the Martians have absolutely no sense of smell, further suggesting their progression along the evolutionary ladder away from idiots, savages, and others in whom the olfactory sense was thought to predominate.

At first glance, Wells's evolutionary predictions suggest his intellectual debt to Allen.[83] But both writers drew upon a widespread Victorian belief that vision and touch were intimately related. Spencer, for instance, compared vision to "anticipatory touch," and emphasized that "visual impressions are habitually followed by tactual ones."[84] Henry Maudsley likewise argued in his 1878 volume *The Physiology of Mind* that civilized Europeans place the heaviest reliance on the sense of sight, but that vision itself developed out of the "primordial sense of touch" and is thus closely connected to it.[85]

Una's behavior conforms to these nineteenth-century views on the connection between sensory perception and cognition, as she must often *touch* photos before she can fully comprehend their content. When questioned by the Scotland Yard inspector, Una precisely recalls the visual details of the crime scene, but says that she "can't remember" the events themselves and feels mentally "blank, blank, as usual" (14). But when the policeman hands her the photo, Una relates, "I took it, all on fire. The sight almost made me turn sick with horror. To my awe and amazement, it was indeed the very scene I remembered so well" (16). Una's simultaneous visual and tactile contact with the photo conjures up visceral emotional memories connected with the scene that has "burned itself onto [her] brain" (21). Holding the photograph, Una instantly understands its emotional context, even if she cannot yet recall what happened before and after. Later, upon seeing a photograph of her real father for the first time, Una "seized the faded photograph and pressed it to [her] lips," *after* which she exclaims, "oh, I know him … It's my father!" (100). In both cases, touching a photograph helps Una to intellectually process the information it contains.

Intriguingly, the abovementioned crime scene photograph that Una "grasps" – both figuratively and literally – depicts the back and hand of a fleeing individual whom Una thinks is her father's murderer. Una tries to identify the murderer by finding the owner of the "soft and delicate hand, very white and womanlike" shown in the photograph (33). This search ultimately leads her to Dr. Ivor, whose womanlike hands belie his otherwise masculine features. Yet because "the hand of the Moores" is a "family peculiarity" among Una's mother's relations, including the doctor, Una first suspects her Aunt Emma and her cousin Minnie of the crime, solely based on their hand shape (48). The resulting confusion calls into question the truth value and uniqueness of photographs, particularly as used in criminal investigations.[86] Moreover, Allen subtly suggests a neural connection between the senses of sight and touch when Una's eyes meet these curiously disembodied photographic hands.

*Recalled to Life* provides an unusually literal representation of photographic memory that posits the eye as camera and the brain as gallery of visual images. The novella embodies Allen's scientific convictions about the primacy of the visual and tactile senses in human beings. But behind Allen's intricate visual metaphors lurks a sense of the insufficiency of photographic truth, since an elaborate sequence of pictures – rather than an individual photograph – provides the key to the solution of Una's mystery. Furthermore, the visual data of a photograph occasionally must be supplemented by tactile stimulation – touching the photo – in order for Una to conjure up the corresponding memory or emotional response. If Una is a camera, she is a seriously flawed one whose disconnected mental images require the objective wisdom of a "man of science" to achieve full coherence.

## GOTHIC TECHNOLOGIES AND TRAUMATIC SHOCK

So far, I have suggested that Allen's facile equation of woman and camera stands up to the scrutiny of late-Victorian science, particularly physiological studies of vision and memory. But when set against the broader perspective of photographic history and early theories of traumatic shock, the equation of brain and machine becomes infinitely more troubled. From here on, we must leave the secure ground of scientific positivism and enter the murkier territory of subjective psychological experience.

And no late-Victorian genre was better adapted to exploring the intricacies of subjective psychological states than the Gothic.

At first glance, Allen's novel appears to be a straightforward detective story. But recent critics have emphasized the ways in which detective fiction overlaps with sensation fiction and the Gothic.[87] Moreover, certain plot elements of *Recalled to Life* suggest the novella's affinities with these two "underworld" genres. For instance, *Recalled to Life* resembles sensation fiction of the 1860s and 1870s in its emphasis on murder, mistaken identities, illegitimacy, madness, and other dangers lurking within the respectable Victorian home. Like Franklin Blake of Wilkie Collins's *The Moonstone* (1868), Una is simultaneously a detective and unwitting criminal.[88] Una's amnesia also has parallels in sensation fiction, in which, as Nicholas Dames suggests, "the process of 'being shocked' by new information shades into the process of 'going into shock.'"[89] For characters in sensation novels as well as for readers of such fictions, temporary amnesia often results from overwhelming exposure to terrifying or scandalous events.[90]

Curiously, catatonic or amnesiac women feature in two Sherlock Holmes stories, *The Crooked Man* (1893) and *The Dancing Men* (1903), where these helpless females are key witnesses and suspects in murders. These examples demonstrate the blurry boundary between the so-called "twin" genres of sensation and detective fiction.[91] Like sensation novels, which often focused on mysterious women harboring disturbing family secrets, much detective fiction of the 1890s featured alluring female detectives and criminals. Joseph Kestner has described how a subgenre featuring female detectives flourished in the wake of Holmes's popularity. Successful Victorian authors like Allen, Richard Marsh, and Mary Elizabeth Braddon penned novels featuring female sleuths.[92] Although most of these popular fictions are now forgotten, their collective influence can be seen in later female detective series, such as the Nancy Drew novels of the 1930s.

*Recalled to Life* also resembles late-Victorian crime fictions featuring glamorous female miscreants.[93] Though Una remains ignorant of her crime until the novella's conclusion, she could still justly be considered an example of what Miller has called the "New Woman Criminal," an attractive female vigilante who frequently appears in detective fiction of the 1880s and 1890s and typically evades capture. The most famous example may be cross-dressing adventuress Irene Adler of Conan Doyle's "A Scandal in Bohemia." Readers enjoyed living vicariously through these female criminals, whose social transgressions stretched the boundaries of *fin-de-siècle* gender roles.[94]

Yet, unlike the competent female detectives and carefree lawbreakers described by Kestner and Miller, Una's compromised psychological state makes her an unlikely champion of women's newfound freedoms. During her post-traumatic "second state," Una is as helpless as a newborn, becoming almost literally "a baby in arms again" (5). For several years following her father's murder, Una is rendered speechless by visual recollections of the trauma she describes simply as "the Picture" or "the Horror":

> This was the Picture as, for many long months, it presented itself incessantly to my startled brain, by day and by night, awake or asleep, in colours more distinct than words can possibly paint them. I saw myself standing in a large, square room … on one side stood a table, and on the table a box. A flash of light rendered the whole scene visible … Half-way to the door, I stood and looked in horror at the sight revealed before my eyes by that sudden flash. A man lay dead in a little pool of blood that gurgled in short jets from a wound on his left breast. I didn't even know at the moment the man was my father; though slowly, afterward, by the concurrent testimony of others, I learnt to call him so. (5)

The remainder of the book relates Una's attempt to deconstruct this overwhelming image that had "photographed itself vividly" on her "mind's eye" (5). The box turns out to be a camera; the flash of light results from an early form of flash photography. But since Una cannot recall what occurred before or after "the Picture," the scene remains for her decontextualized, unfamiliar, and uncanny.

Una's intrusive, overwhelming recollections of "the Horror" lend the novella a Gothic flavor. Not only are her memories psychologically distressing, but their expressly visual character arguably intensifies the aura of mystery surrounding them. Whereas Paley and Darwin viewed the human eye as a "perfect" structure, other nineteenth-century scientists such as Kühne, Böll, and Hermann von Helmholtz studied imperfections in human vision, including blind spots, floaters, retinal after-images, and other visual anomalies. These curious phenomena suggested the subjective, unreliable nature of human visual perception, even when the eye and brain were in a healthy state and the individual under study was not prone to visual hallucinations. Allen knew of these studies, including Helmholtz's *Physiological Optics* (1856–66) and Kühne's studies on frog retinas, both of which he mentioned in *Physiological Aesthetics* and *The Colour Sense*.[95] Allen found retinal after-images particularly fascinating, discussing them at length in *Physiological Aesthetics* and elsewhere.[96]

Studies examining subjective phenomena of vision subtly evoked a connection between the visual and the Gothic. As Andrea Goulet points out, the optical treatises of Helmholtz and others tended to figure "the eyeball – or, more specifically, the retina – as a particularly troubled site of human epistemology."[97] Nineteenth-century fiction followed suit. For example, in Charlotte Perkins Gilman's "The Yellow Wallpaper" (1892), normal vision gradually shades into psychosis as the heroine's perception of the wallpaper changes "as the light changes."[98] Although nineteenth-century physiologists confirmed that painted paper does appear to change as the light shifts, this ordinary visual phenomenon becomes "a gateway toward possible pathology" when the protagonist's husband views it as a psychotic symptom.[99] The fallibility of human vision likewise contributes to the sense of mystery pervading *Recalled to Life*. Like Gilman's heroine, Una's visual memory repeatedly comes under scrutiny from male authority figures, including her future husband and the Scotland Yard inspector.

The fact that Una's memories are suppressed or hidden likewise intensifies the novella's Gothic undertones. Nils Clausson reminds us that *fin-de-siècle* Gothic tales often feature anxieties about "a long-dormant or suppressed past re-emerging in the present."[100] In the course of confrontation with a "haunting and/or imprisoning past," old sins re-emerge and must be expiated.[101] Detective fictions likewise concern past crimes that surface unexpectedly and must be atoned for. But the resolutions offered by detective fictions are often unsatisfactory, Clausson explains: "The *fin-de-siècle* detective story, like its close literary cousin the *fin-de-siècle* Gothic tale, operates both to create mystery and then to give the illusion – but only the illusion – of solving it."[102]

Just as human vision is liable to distortion and manipulation, photographs can likewise be viewed as illusions rather than reliable records of past events. Daniel Novak describes how frequently Victorian photographs were retouched or pasted together for artistic effects (similar to the modern practice of digitally altering photographs).[103] Nonetheless, Novak writes, "critics have most often argued that the Victorians trusted the objectivity of photography."[104] Some critics have even viewed photographs as somehow more real than the objects they depict, a view famously articulated in André Bazin's essay, "The Ontology of the Photographic Image" (1945): "The photographic image is the object itself, the object freed from the conditions of space and time that govern it."[105] Miller argues that, like Bazin, Sherlock Holmes "treats photography … as a fetishized or idealized form of reality and an utterly transparent window into history."[106] A similar view surfaces in *Recalled to Life*. Echoing the language of instantaneous photographers themselves, who often described their images as "from life" or "from Nature," Una reflects that a photograph "couldn't lie: I knew it must be the very handiwork of unerring Nature" (93).[107]

But Una's equation of photography and truth is qualified within the novella as a whole, in which Allen suggests that knowing a photograph's context is crucial to understanding its contents. For instance, Una's misidentification of the murderer depends on the angle of view from which her mental "Picture" was "taken," as well as her inability to contextualize the image. She discovers her mistake when Dr. Ivor presents her with a missing portion of the acmegraphic sequence taken at the time of the murder. This "instantaneous photograph … represented a scene just before the one the Inspector gave me. And there, in its midst, I saw myself as a girl, with a pistol in my hand" (92). In *Recalled to Life*, therefore, even the most accurate photos are liable to misinterpretation if viewed out of context.

In the Victorian era, as now, the type of photographic technology used could affect a photograph's accuracy. Early Victorian photographic techniques required long exposures, and photographic sittings often lasted for several hours. As a result, early photographs sometimes presented "impossible" images, such as spectral human forms created by incomplete exposures. Photographers often misrepresented such incomplete exposures as pictures of ghosts or spirits. Allen's friend Conan Doyle was notoriously gullible about such images, which bolstered his belief in the existence of ghosts and fairies.[108]

Decreasing the length of photographic exposures arguably increased the scientific accuracy of the photographic process. In the early 1850s, photographer William Henry Fox Talbot experimented with high-speed photography using his new amphitype process. This process involved making rapid exposures using an electric spark to provide extra lighting – the earliest form of flash photography. Talbot's methods were imitated by many photographers hoping to capture so-called "instantaneous" images using rapid exposures and artificial lighting.[109] Until the 1888 invention of George Eastman's portable Kodak camera, such technology was generally available only to professional photographers.

Allen situated his Gothic detective tale at a transitional moment in the history of Victorian photography, shortly after the invention of the Kodak camera and just before the invention of cinema. The 1870s, 1880s, and 1890s were also the heyday of chronophotography, whose methods Allen faithfully depicts in *Recalled to Life*. In the 1870s, chronophotographer

Eadweard Muybridge famously photographed a trotting horse "to settle a long-standing controversy among racing men as to whether a trotting horse had all four hooves off the ground at any point."[110] In his celebrated book *Animal Locomotion* (1887), Muybridge expanded his subject matter to include sequential pictures of other animals and nude human beings in rapid motion. Muybridge's photos depicted stages of swift movements occurring "too rapidly to be seen with the naked eye," effectively defamiliarizing everyday activities such as walking, galloping, etc. (see Figure 4).[111] Like Muybridge, Vivian Callingham captures sequential stages of rapid motion using his acmegraphic process, and his favorite photographic subjects ("men running and horses trotting") mirror those preferred by Muybridge (50).

Capturing such rapid movements on film simultaneously clarified the processes of human and animal locomotion and mystified viewers unused to seeing fleeting actions so precisely recorded. Muybridge's photography thus highlights a phenomenon Walter Benjamin has described as "the optical unconscious," in which the camera reveals hidden aspects of movement that normally register below conscious awareness: "While it is possible to give an account of how people walk, if only in the most inexact way, all the same we know nothing definite of the positions involved in the fraction of a second when the step is taken. Photography, however, with its time lapses, enlargements, etc. makes such knowledge possible."[112] This statement highlights the Gothic potential of chronophotography and related technologies, which could alienate viewers from everyday actions. At the same time, scientists like Francis Galton felt drawn to the apparent objectivity of the medium as well as its potential to record movements too subtle to be perceived without mechanical assistance.[113] Like detective fiction, then, chronophotography was an ostensibly objective medium with quasi-Gothic undertones.

These improvements in high-speed photography helped pave the way for the invention of cinematic technologies in the 1890s, as did the zoopraxiscope movie projector invented by Muybridge in 1879. Muybridge's zoopraxiscope (literally, "animal action viewing device") reanimated his still photos of animals and athletes in motion, effectively creating short films out of a series of sequential photographs. Muybridge used this device both to entertain viewers and to prove that his instantaneous photographs were authentic and unretouched. While Muybridge's still photos had the effect of defamiliarizing common movements, reanimating these images using the zoopraxiscope allowed audiences to recognize everyday actions

Figure 4  Eadweard Muybridge, photograph of a trotting horse (1878).

like a horse galloping or an athlete clearing a hurdle.[114] On account of this invention, historians such as Gordon Hendricks have hailed Muybridge as "the father of modern cinema," though others dispute this claim.[115] Certainly, Muybridge's chronophotography and his zoopraxiscope were important forerunners of inventions such as Edison's Kinetoscope and the Lumière Cinématographe movie projector (invented in 1891 and 1895, respectively).[116]

*Recalled to Life* presents remarkably detailed depictions of flash photography and chronophotography, both of which play a role in Una's trauma and recovery. While flash photography initiates Una's trauma and her "second state," chronophotography facilitates healing by helping her patch together her disjointed memories. Kate Flint argues that in Grant Allen's work, as elsewhere in Victorian writing, flash photography may be obscurative and disruptive rather than immediately illuminating.[117] Allen wrote in *Physiological Aesthetics* that "any violent and sudden sense stimulant, such as the roar of a cannon, a flash of lightning, a shooting pain, or an unexpected shaking … forcibly interrupts the regular course of consciousness."[118] While this passage does not explicitly mention flash photography, *Recalled to Life* includes frequent comparisons between lightning flashes and flash photography techniques.[119] At the moment of her father's murder, for instance, Una describes the "flash of light" from her father's acmegraphic device as "rather like lightning, so quick it was, and clear, and short-lived, and terrible" (5).

The "short-lived," "terrible" flash from Vivian Callingham's camera produces a physical shock that "obliterate[s] whole tracts" in Una's memory (45). In describing Una's trauma and its aftermath, Allen combined several late-Victorian views of human consciousness. Ribot speculated that "the mechanism of consciousness is comparable to that of vision."[120] He suggested that consciousness consists of a series of "evanescent trace[s]" akin to retinal after-images that gradually fade.[121] This view emphasized the ephemeral, incomplete nature of human recollection. Other scientific writers, including Herbert Spencer and Alexander Bain, speculated that consciousness consisted of a sequence of minute electrical shocks.[122] Spencer, for instance, proposed that "all existence distinguished as subjective, is resolvable into units of consciousness similar in nature to those which we know as nervous shocks."[123] Along the same lines, many Victorians conceived of a particularly strong emotion as a jolt of electricity.[124] These arguments were based on the emergent awareness that nerve signals were electrical in nature, a fact firmly established during the late-Victorian period.[125] Allen brought together these strains of thought in *Recalled to Life*, where he implied that too much electrical stimulation – such as the "electric spark" then used in flash photography – could interrupt pictorial consciousness (and memory itself) by overloading the delicate human nervous system.

In depicting Una's illness, Allen also drew from pre-Freudian trauma theories that figured a traumatic event as an actual injury to the brain and spinal column. Such theories originated during a spate of railway disasters in the 1850s and 1860s. Doctors wondered why some victims of these accidents exhibited psychological distress and cognitive impairment in the absence of obvious physical injury. One such patient was Charles Dickens, who suffered from what we would now call post-traumatic stress following an 1865 railway accident. His symptoms included shaking, feelings of panic, and the temporary loss of his voice (reminiscent of Una's aphasia).[126] In such cases, physicians often argued that patients' symptoms were caused by physical, albeit undetectable, injuries to the brain or spinal column.[127]

During her railway journey through Canada, Una is involved in a train collision that leaves her mildly injured and "very much shaken," an incident that seems calculated to remind us of the source of much mid-Victorian trauma theory (85). Yet, as Jill Matus suggests, late-Victorian studies of "psychic shock" (including "railway spine") did not tend to focus on memory disruption so much as emotional disturbances of other kinds – anxiety, depression, sleeplessness, and so forth.[128] In focusing on amnesia and aphasia resulting from traumatic shock, Allen apparently anticipated later trauma theories that would focus on cognitive and memory disruption as the cardinal signals of trauma.

While it is Freud who is best remembered for connecting trauma and memory disruption in his 1895 collaboration with Josef Breuer, *Studies on Hysteria*, he was not the first to describe cognitive abnormalities following traumatic experiences. In *Diseases of Memory*, for instance, Ribot described hypermnesia (unusually intense or accurate recall of events) resulting from illness, "maniacal excitation," or exposure to extreme dangers. Ribot mentions persons saved from drowning "who agree that at the moment of asphyxia they seemed to see their entire lives unrolled before them in the minutest incidents." One such individual compared his experience to another kind of popular photography practiced by Muybridge, stating that he had observed "a kind of panoramic picture of his entire existence" as death approached.[129]

Ribot also recorded cases of amnesia during which some aspects of patients' memory improved. One twenty-four-year-old female amnesiac was particularly remarkable in this regard: "her memory … so entirely lost as far as regarded previous knowledge, was soon found to be most acute and retentive with respect to everything she saw or heard subsequently to her disorder."[130] Similarly, while Una has forgotten events prior to her father's murder, her skill at amateur detection and the rapid progress of her re-education suggest that her post-traumatic memory is exceptionally strong. Allen apparently embraced Ribot's notion of memory as "a vision in time" that could be disrupted or intensified by psychological or physiological shock.[131]

Whereas flash photography contributes to Una's mental decline by interrupting and overwhelming normal consciousness, chronophotography ultimately aids in her recovery. Allen's detailed knowledge of chronophotography – both its methods and its most famous practitioners – becomes evident early in the novella. Not only do Callingham's photographic techniques mirror those of Muybridge, but he also shares the famous photographer's turbulent personality. Muybridge was well known for his scandalous personal life. In 1874, he murdered his wife's lover when he discovered their affair and the illegitimacy of his only child. He was acquitted on the ground of "justifiable homicide" in 1875, after which he fled to South America until the controversy subsided.[132] Intriguingly, some have speculated that Muybridge's erratic behavior was caused by a brain injury resulting from a stagecoach accident, though this theory is impossible to verify posthumously.[133] Callingham resembles Muybridge in his long white beard, his attempted slaying of Una's real father, and his violent temper.

Alternatively, Callingham may be based on French chronophotographer Etienne-Jules Marey, who in 1882 pioneered the "photographic gun," the first camera to make sequential images using a single lens. Behind this lens was a "photographically sensitive disc that spun as an exposure was made." The advantage of the photographic gun, Phillip Prodger explains, was that "it provided a single perspective from which the subject could be seen … the effect was that of a single eye surveying events." By contrast, Muybridge had used multiple cameras triggered by trip wires or clockwork mechanism to create the photographs in *Animal Locomotion*.[134] Vivian Callingham's "apparatus that let sensitive plates revolve one after another opposite the lens of a camera" sounds similar to Marey's photographic gun, although Callingham's electric flash is triggered by a "clockwork mechanism" such as Muybridge used (17). Callingham's acmegraph appears to be Allen's creative amalgam of available chronophotographic methods.

Allen draws striking parallels between Vivian Callingham's chronophotography and Una's photographic memory. In her amnesiac state, Una remembers isolated images divorced from their context – her memories are like individual pictures from chronophotographic sequences. Una's still mental "Picture" of "the Horror" strikes her as unfamiliar, even uncanny, just as the still photographs from Muybridge's trotting horse sequence originally appeared unreal or impossible to viewers. Una's memory of the crime scene is of "so much isolated and unrelated fact, without connection of any sort to link it to the events that preceded or followed it" (93).

As her memories return, however, Una's brain operates like Muybridge's zoopraxiscope, which familiarized still images by reinserting them into the sequences of which they formed a part. As Frank writes, nineteenth-century literary detectives must often "explain … a fact or an event by placing it within a chronological series."[135] But Una's role as detective goes beyond deciphering the events leading to her father's murder. Like Muybridge's zoopraxiscope, Una's sequential ordering of photographs provokes a shock of recognition that proves the authenticity of each image. Prodger explains of Muybridge's invention: "Skeptics might not have believed in the veracity of [Muybridge's] photographs on their own, but when they saw them projected in motion, they saw something they recognized and could mentally verify."[136] Similarly, when Una finally views the photo of herself as murderer that completes her father's acmegraphic sequence, she "remember[s] it all as something that once really occurred to me" (93).

Una's recognition of an isolated memory as part of a sequence can be simultaneously distressing and cathartic, and often affects her physically – as one might expect from late-Victorian theories of trauma as spinal or cranial injury. For instance, when Una first recognizes Dr. Ivor in one of her father's photographs, she experiences a "cold thrill" of recognition (50). Similarly, when she finally views the picture of herself holding the smoking pistol, she falls back "with a deadly shriek of horror" (92). Una is not only horrified to discover herself a murderer, but also overwhelmed by the rapidity of her returning memory: "I remembered it so, now … I remembered having stood, with the pistol in my hand, pointing it straight at the breast of the man with the long white beard whom they called my father" (93).

Una's visceral responses to her returning memories occasionally bring to mind later, psychoanalytic discussions of traumatic shock; take, for instance, Breuer's famous patient, Anna O., who also experienced aphasia and amnesia following her father's death.[137] But in fact, Una's behavior conforms to earlier nineteenth-century thinking about the relationship between emotion and cognition. Una embodies Spencer's idea that "every cognition is a recognition," as she cannot truly understand a past event until she remembers it, visually and viscerally.[138] She likewise demonstrates the related Spencerian principle that "no act of cognition can be *absolutely* free of emotion," since Una's flashes of insight always provoke concomitant emotional and physical responses.[139]

In a similar vein, Ribot studied cases of progressive amnesia (such as senile dementia) and emphasized that the last mental functions to deteriorate were emotions. This fact suggested the primacy of emotion, and the relatively ephemeral character of acquired knowledge.[140] In cases of temporary amnesia, Ribot suggested that the first memories to return were emotional ones, followed by simple concepts learned in childhood. These principles are borne out by Una's recovery. For instance, when Una returns to her hometown for the first time following the murder, she finds that "the appropriate emotion" associated with certain persons and places "seemed easier to rouse … than the intellectual memory" (30).

Ribot's insistence on the mechanical nature of consciousness may help to explain why Allen was particularly attentive to his theories regarding amnesia. As a man's memories and behaviors become increasingly ingrained, Ribot wrote, "all his acts will be entirely automatic. Shallow and commonplace minds realize this hypothesis to a certain extent … they become mere machines."[141] But unlike Allen, Ribot objected to inorganic metaphors describing memory as a photographic imprint, instead insisting that "the bases of memory must be looked for in the properties of organic matter, and nowhere else."[142] By taking the idea of a mechanical consciousness so literally, Allen departed from Ribot and from the thinking of all but the most radical Victorian scientists.

## CONCLUSION

Grant Allen devoted himself to a brand of scientific rationalism that made no room for the occult, the supernatural, or the inexplicable. This empirical viewpoint is obvious not only in Allen's scientific writing, but also in his fiction. Allen's detective novels suggest his commitment to scientific logic, while his interest in photography implies his desire for a permanent, infallible record of the past. Yet just as the detective genre subtly shades into the Gothic, the photograph can be an illusion rather than a truthful representation of reality. Moreover, in *Recalled to Life*, the fallibility of human vision and human memory perpetually threaten to undermine the creation of a logical, sequential narrative.

In his creation of a "human camera" who dramatizes the mechanical workings of the mind, Allen conspicuously fails. This failure is not due to lack of scientific accuracy on his part, nor to any misunderstanding of photographic technology. Rather, Allen's talent for Gothic description undermines his commitment to mechanical precision. Moreover, his psychologically accurate rendering of his heroine's emotional state lends her unexpected depth of character. The reader is left with the haunting impression of "the Horror" that initiates Una's mental decline and foregrounds her crushing psychological vulnerability. Thus, even though Allen appealed to a mechanical model of consciousness rooted in late-Victorian physiology, he inadvertently foreshadowed Freud's dynamic understanding of trauma as a break in narrative consciousness.

Of the writers in this study, Allen would seem to be the most committed to cerebral localization and the most comfortable with biological determinism in all of its forms. Like Huxley, Clifford, or Hodgson, Allen placed himself at the forefront of the "Determinist army" that gradually turned psychology from a science of the soul into a science of the brain, reducing the brain to mechanism along the way.[143] But the seepage of the Gothic into his rational detective novella, not to mention the runaway humanity of his human camera, suggests that Allen never fully embraced cerebral localization and its materialist consequences (despite his statements to the contrary). *Recalled to Life* thus shows that even the scientific vanguard could not face down the specter of the human machine conjured up by late-Victorian neurology.

# PART III

# *Visionaries*

CHAPTER 4

# *H. G. Wells and the evolution of the mad scientist*

In 1893, H. G. Wells's article "The Man of the Year Million" dramatically predicted the distant evolutionary future of mankind:

> The descendants of man will nourish themselves by immersion in nutritive fluid. They will have enormous brains, liquid, soulful eyes, and large hands, on which they will hop. No craggy nose will they have, no vestigial ears; their mouths will be a small, perfectly round aperture, unanimal, like the evening star. Their whole muscular system will be shrivelled to nothing, a dangling pendant to their minds.[1]

The editors at *Punch* evidently found this prediction hilarious, publishing a poem and accompanying sketch ridiculing Wells's lopsided future humans (see Figure 5). But not everyone was laughing.

As ridiculous as Wells's bodiless, large-headed "human tadpoles" may seem, they were based on the most rigorous evolutionary science of their day.[2] Wells, a lower-middle-class academic prodigy, received a prestigious government scholarship to attend the Normal School of Science in South Kensington (later absorbed into the University of London). Though Wells left South Kensington in 1887 without earning his degree, he was greatly inspired by his biology teacher, famed physiologist Thomas Henry Huxley. Wells absorbed Huxley's pessimism about the direction of evolution, particularly his emphasis on the inherent brutality of natural selection.

Huxley's pessimism surfaces in Wells's dystopian scientific romances, which imaginatively probe the consequences of evolutionary theory run amok.[3] Beginning with the mad scientists of *The Island of Doctor Moreau* (1896) and *The Invisible Man* (1897), and continuing with the extraterrestrials of *The War of the Worlds* (1898) and *The First Men in the Moon* (1901), Wells depicted brains becoming steadily larger and more powerful as bodies grow smaller and more useless, emotions increasingly muted, and conscience all but silenced. Wells's nightmarish vision of the massively over-evolved brain unites these four works, as ruthlessly intellectual

## 1,000,000 A.D.

[“The descendants of man will nourish themselves by immersion in nutritive fluid. They will have enormous brains, liquid, soulful eyes, and large hands, on which they will hop. No craggy nose will they have, no vestigial ears; their mouths will be a small, perfectly round aperture, unanimal, like the evening star. Their whole muscular system will be shrivelled to nothing, a dangling pendant to their minds.”—*Pall Mall Gazette, abridged.*]

WHAT, a million years hence,
will become of the *Genus*
*Humanum*, is truly a
question vexed;
At that epoch, however, *one*
prophet has seen us
Resemble the sketch
annexed.

For as Man undergoes
Evolution ruthless,
His skull will grow “dome-
like, bald, terete”;
And his mouth will be jawless, gumless, toothless—
No more will he drink or eat!

He will soak in a crystalline bath of pepsine,
(No ROBERT will then have survived, to wait,)
And he'll hop on his hands as his food he steps in—
A quasi-cherubic gait!

No longer the land or the sea he'll furrow;
The world will be withered, ice-cold, dead
As the chill of Eternity grows, he'll burrow
Far down underground instead.

If the *Pall Mall Gazette* has thus been giving
A forecast correct of this change immense,
Our stars we may thank, then, that *we* shan't be living
A million years from hence!

Figure 5 *Punch* cartoon “1,000,000 A.D.”

scientists like Moreau and Griffin morph into the amoral, top-heavy Martians and lunar inhabitants.

Wells's malevolent mad scientists and extraterrestrials owe an intellectual debt not only to Huxley, but also to the clinical association between genius and insanity that developed in the mid-nineteenth century. Numerous scientists, including Jacques Moreau, Francis Galton, and Cesare Lombroso, suggested that mankind had evolved larger brains at the expense of muscular strength, reproductive capacity, and moral sensibility. Perhaps surprisingly, their arguments proceeded from Lamarckian rather than Darwinian evolutionary logic. French zoologist Jean-Baptiste Lamarck and his followers maintained that "hypertrophy" or excessive growth of any given body part – in this case, the brain – was always compensated by atrophy of other body parts. According to Lamarck's first law, expounded in *Zoological Philosophy* (1809), "Use of any organ gradually strengthens, develops and enlarges that organ … while the permanent disuse of any organ imperceptibly weakens and deteriorates it, and progressively diminishes its functional capacity, until it finally disappears."[4] Lamarck's second law, meanwhile, stated that changes brought about by use or disuse of any organs would be inherited by the next generation. This logic still resonated with many *fin-de-siècle* scientists, for whom Darwin's theory of natural selection had not entirely supplanted the earlier Lamarckian concept of inheritance of acquired characteristics.[5] According to such reasoning, the genius who overused his brain could expect his cerebrum to expand while the rest of his body wasted away and his moral sensibilities languished. The stereotypical mad scientist familiar to readers of Mary Shelley's *Frankenstein* (1818) or Robert Louis Stevenson's *Strange Case of Dr. Jekyll and Mr. Hyde* (1886) was thus explained by evolutionary biology.

The association between insanity and genius was widely regarded as "scientific fact" by the late-Victorian period, thanks in part to Lamarckian evolutionary biology.[6] But late-Victorian cerebral localization theory and its pseudoscientific predecessor, phrenology, likewise played a role in this development. As a child, phrenological pioneer Franz Joseph Gall had observed that his classmates with excellent verbal memories tended to have bulging "cow's eyes."[7] He later speculated that these bulging eyes were caused by unusually large frontal convolutions that affected the development of the orbital cavities. Gall proposed that language and all other "higher" intellectual faculties must be localized in the frontal lobes.[8] Phrenologists thus viewed individuals with large skulls and large foreheads as intellectually gifted.

Even at mid century, after phrenology had fallen out of favor in the scientific community, many researchers remained convinced that the frontal lobes must contain the seats of higher faculties. Jean-Baptiste Bouillard and Paul Broca, for instance, both supported the idea that language was localized in the brain's frontal regions. In 1861, Broca performed an autopsy on a deceased patient, M. Leborgne, who had suffered from severe speech difficulties. Based on the injuries he observed, Broca located the faculty of speech in the second or third frontal convolution of the left hemisphere.[9] From these findings and other cases he had seen, Broca felt confident in stating: "The most noble cerebral faculties, those that constitute understanding properly speaking, such as judgment, reflection, the faculties of comparison, and abstraction, have their seat in the frontal convolutions, whereas the temporal, parietal, and occipital lobe convolutions [located in the middle and back of the brain] are appropriate for the feelings, penchants, and passions."[10] Once a highly respected researcher like Broca had voiced these opinions, scientists throughout Europe exhibited marked preference for the frontal region of the brain over the middle and back portions, similar to the way in which they validated the supposedly more "civilized" left hemisphere over the "primitive" right brain (as discussed in Chapter 1).

Frontal-lobe superiority became established through avant-garde cerebral localization research, but was later adopted as a central tenet of that less-respectable Victorian enterprise, craniometry. Stephen J. Gould has described how scientists such as Broca, Huxley, and Galton calculated intelligence on the basis of overall brain size in combination with large frontal lobes and more numerous cerebral convolutions, all of which were thought to be signs of great intellect.[11] Cranial measurements were thought to be perhaps the most reliable and "objective" indicator of intelligence prior to the advent of IQ tests (the first versions of which were developed around 1905 by Alfred Binet).[12] While Victorians considered craniometry to be cutting-edge science, it appears less impressive in hindsight, especially when one considers the culturally biased uses to which it was put. Many craniometrists, Broca included, did little more than validate white male superiority on the basis of skull size and shape, massaging or misinterpreting data that did not support their a priori conclusions.[13]

Wells's early scientific romances, with their large-headed, big-eyed, amoral mad scientists and aliens, demonstrate the wide-ranging influence of these two dubious scientific trends – craniometry, with its emphasis on the shape and size of the brain, and Lamarckian evolutionary biology, which presupposed rapid, heritable changes in brain development. Wells also meditated on these theories in scientific writings like "The Man of the Year Million," and in his biology textbook *The Science of Life* (1929), which he co-authored with his son, George Philip Wells, and Julian Huxley, the grandson of his mentor. In this textbook, the authors explained how the human bone structure supposedly presented an obstacle to human intellectual progress: "In the size of the female human pelvis, for example, there may be a limiting condition to the size of human babies' heads, so to the expansion of the human brain, and so to the elaboration of the human mind."[14] Although one could speculate about the evolutionary possibilities that might result if the skull and pelvis disappeared, such ruminations were open-ended and impossible to support, since they did not easily lend themselves to scientific experimentation.

Wells's fiction was the place where he really put late-Victorian theories about unchecked Lamarckian brain development to the test, doing away with artificial and natural barriers to brain evolution. In 1933, Wells described his early "scientific romances" as imaginative vehicles for testing scientific hypotheses.[15] In works like *The Invisible Man* and *The Time Machine* (1895), Wells concentrated on one hypothetical proposition, such as "how would you feel and what might not happen to you … if you became invisible?"[16] Wells then isolated this imaginative "what if" question and made all other aspects of the story as realistic and commonplace as possible:

> For the writer of fantastic stories to help the reader to play the game properly, he must help him in every possible unobtrusive way to *domesticate* the impossible hypothesis. He must trick him into an unwary concession to some plausible assumption and get on with his story while the illusion holds … as soon as the magic trick has been done the whole business of the fantasy writer is to keep everything else human and real. Touches of prosaic detail are imperative and a rigorous adherence to the hypothesis. Any *extra* fantasy outside the cardinal assumption immediately gives a touch of irresponsible silliness to the invention.[17]

This elaborate manner of crafting a romance strikingly resembles the scientific method, wherein one variable is tested against a series of controls. The likeness makes sense, given Wells's extensive training in experimental physiology under Huxley in 1884.[18]

The results of Wells's fictional "experiments," however, were more Gothic than rational. Wells envisioned dystopian near-future scenarios in which human mad scientists and giant brains from outer space blight any hopes for human progress. What made Wells's visions especially chilling was their plausibility according to *fin-de-siècle* craniometry and evolutionary biology. By dispensing with skulls and human forms of reproduction, Wells's Martians and lunar inhabitants achieve intellectual superiority over human beings, developing enormous brains that dwarf their spindly bodies. But these aliens also strikingly demonstrate the supposed kinship between high intelligence, insanity, and amorality described by late-Victorian scientists.

Time has not been kind to the scientific theories Wells tested in his imaginative romances. Lamarckism is generally viewed as a relic of nineteenth-century scientific thought, while craniometry is reviled as a form of cultural imperialism masquerading as science.[19] Yet the prototypical aliens of modern science fiction, television, and cinema nearly always match Wells's description, boasting large heads, big eyes, and small bodies. Wells's Gothic imagination has proved more permanent than the science upon which he based his extraordinary visions.

Nonetheless, we cannot fully appreciate Wells without understanding the now-forgotten scientific theories upon which he based his visionary tales. Thus, the following section will explore the clinical association between genius and insanity prevalent during the late nineteenth century, along with the dubious scientific hypotheses that undergirded it. Along the way, the reader will rediscover a number of obscure scientific figures, like Scottish philosopher John Ferguson Nisbet, whom Wells curiously described as "a brain, a resolute interrogative brain."[20] Nisbet was one of many late-Victorian thinkers who felt convinced that genius always bespoke serious moral and psychological deficiencies, and that large brains possessed some inevitable drawbacks.

## GENIUS AND INSANITY

Because Nisbet's work synthesized earlier scholarship about genius and insanity and proved especially influential to Wells, his 1891 volume *The Insanity of Genius* will serve as an excellent ideological (if not chronological) point of departure. In this volume, Nisbet declared: "genius, insanity, idiocy, scrofula, rickets, gout, consumption, and the other members of the neuropathic family of disorders" reveal "want of equilibrium in the nervous system."[21] According to Nisbet's reviewer in *Mind*, the author assumed that "superiority and inferiority to the average are to be classed together as deviations from the normal" and that abnormality was necessarily pathological.[22] Nisbet also argued that genius was co-morbid with various types of mental illness. More importantly, he defined genius itself as a kind of hereditary, degenerate brain condition symptomatic of "nerve disorder" that "runs in the blood."[23]

Strange as Nisbet's conclusions may seem, he insisted that his inquiry into genius and insanity was based on cutting-edge science. Nisbet explained: "The results of modern research affecting most intimately the question of genius are, first, the localization of the functions of the brain, and, secondly, the established kinship of an extensive group of brain and nerve disorders."[24] Citing the work of cerebral localizationists such as David Ferrier, Gustav Fritsch, and Eduard Hitzig, Nisbet concluded that genius and insanity both consist in "instability or a want of equilibrium in the nervous system" occurring when certain parts of the brain develop at the expense of others.[25] Nisbet was particularly interested in the even, balanced development of the frontal lobes as a basis for mental health, since "Ferrier and others have shown … that the frontal lobes form the substrata of … the higher intellectual processes."[26]

Nisbet undermined his compelling arguments, however, by failing to provide a consistent psychological definition of genius. Like many other writers linking genius with insanity, Nisbet assumed that accomplishment bespoke intellectual aptitude, thus confusing success and fame with innate ability. Despite this flaw, *The Insanity of Genius* was reprinted five times before 1912, perhaps because it articulated a timely and popular idea. Late-nineteenth- and early-twentieth-century British and continental authors produced hundreds of monographs on genius and insanity.[27] The cumulative result was the widespread belief in a "scientific" relationship between genius and mental illness during the late-Victorian era – even though the volumes making this claim (like Nisbet's) were by and large surprisingly unscientific, relying primarily upon anecdotal evidence rather than experimental or statistical data.[28]

Of course, Victorians were not the first to correlate genius and mental illness. This association began with classical authors, notably Plato, Seneca, and Aristotle, who famously declared that "no great genius has ever been without some touch of madness."[29] "Genius" is in fact a Latin word derived from the Greek *ginesthai* ("to be born or created"). In classical pagan tradition, a "genius" referred to the guiding or tutelary spirit allotted to each person at birth. From the Renaissance until the eighteenth century, English authors frequently invoked the older meaning of genius as tutelary spirit. This usage gradually gave way in the nineteenth century to the now familiar definition of genius as superlative intellectual ability (or a person possessing such ability).[30]

Significantly, the period when this linguistic transition occurred – the late eighteenth and early nineteenth centuries – witnessed the birth of several modern myths about genius. Whereas Enlightenment authors described the genius as directly inspired by God, Romantics developed a secularized version that emphasized the artist or poet himself as a godlike figure. Politically, the Romantic genius was a rebel who challenged social hierarchies by suggesting that innate creative powers trump class status. Romantics embraced the classical notion of the creative process as an irrational *furor poeticus*: an "aesthetic response [that] seemed to involve a spell, a rapture, a delirium, a momentary madness."[31] The creative process was also inherently solitary and necessitated a certain degree of suffering for one's art. In England, the Romantic figure of the godlike yet tormented genius was memorably embodied by Lord Byron's self-loathing protagonists, the eponymous hero of Percy Shelley's *Prometheus Unbound* (1820), and Mary Shelley's Victor Frankenstein.

Whereas the Romantics saw genius as a mystical phenomenon beyond the reach of scientific investigation, Victorians brought scientific tools and techniques to bear on superlative ability. Rather than glorifying creative powers, Victorians pathologized genius and upheld the mediocre man as an evolutionary ideal. Michel Foucault has discussed how the impetus toward medical and social conformity reasserted itself as the Romantic cult of individuality waned. Nineteenth-century scientists "embrace[d] a knowledge of *healthy man* … and a definition of the *model man*."[32] This emphasis on normalcy derived partly from Belgian mathematician Adolphe Quételet's influential statistical concept of the "homme moyen" or average man. Quételet's essay *A Treatise on Man and the Development of His Faculties* (1835) presented his so-called "average man" as the central value around which measurements of a human trait are grouped according to the normal distribution.[33] According to this model, all aberrations from the norm could be seen as pathological, including extreme intelligence.

Victorians studying genius and insanity eagerly embraced Quételet's ideas. English eugenicist Francis Galton, a prodigy who could read, write, and conjugate Latin verbs by the age of four, employed Quételet's law of deviation from an average throughout his landmark study *Hereditary Genius* (1869).[34] One of the few scientifically rigorous studies of genius before 1900, Galton's *Hereditary Genius* incorporated intelligence distribution charts that were clearly forerunners of the IQ bell-curve.[35] Since *Hereditary Genius* pre-dated IQ tests by several decades, however, Galton charted measurements of cranial size and shape, exit exam scores at Oxford and Cambridge, and degrees of kinship between acknowledged geniuses and their distinguished relatives, among other supposed indicators of intelligence. Less mathematically inclined authors like Nisbet and Huxley likewise absorbed Quételet's view of normalcy as an index of health. Nisbet argued, for instance, that "[i]t is inevitable that all departures from the mean, in the human species, including those which constitute genius, should be unsound … the greater the genius, the greater the unsoundness."[36] Such apparently democratic sentiments spelled trouble for Victorian geniuses, who suddenly found themselves classed among lunatics and imbeciles at the fringes of statistical charts.

Whereas the Romantics idealized poetic genius, Victorians increasingly focused on scientific prodigies. Cesare Lombroso, professor of psychiatry and forensic medicine at Turin and author of *The Man of Genius* (1864; English translation 1891), defied the common view that "mathematicians are exempt from psychical derangements" by including mathematical greats like Isaac Newton and Blaise Pascal in his list of mad geniuses.[37] Galton, meanwhile, devoted an entire book to scientific precocity, *English Men of Science: Their Nature and Nurture* (1874).

Scientific genius was a timely theme well suited for imaginative literature. Significantly, the rise of the mad scientist as fictional trope coincided with the growth of scientific professions. Roslynn Haynes has described how pejorative stereotypes about scientists intensified following the Industrial Revolution and the biological revolutions of the Victorian era, resulting in a plethora of intriguing literary figures. "The alchemist" is a Faustian character who "reappears at critical times" in the history of science. He obsessively pursues arcane intellectual goals redolent of ideological evil (recently, Haynes notes, such characters are usually biologists). A second stereotype, the "unfeeling scientist" who has "suppressed all human affections in the cause of science," is clearly a legacy of Shelley's Victor Frankenstein.[38] Late-Victorian mad scientists were often composites of these two stereotypes. Moreau is a key example, as are the eponymous protagonist of Stevenson's *Jekyll and Hyde*, Dr. Raymond of Arthur Machen's *The Great God Pan* (1894), and the sinister vivisector Dr. Nathan Benjulia in Wilkie Collins's *Heart and Science* (1883). The first draft of *Moreau*, with its deleted references to *Frankenstein* and its structural resemblance to *Jekyll and Hyde*, suggests that Wells self-consciously situated his novel within this emergent tradition of Gothic mad scientist fiction.[39]

While it is understandable that literary authors might capitalize on lay fears surrounding science and scientists, this does not explain what Victorian scientists stood to gain by doing so. Why would bright men like Galton or Lombroso risk implicating themselves by classifying genius (especially scientific genius) alongside madness? Tellingly, biologists and psychologists seem suspiciously underrepresented in Victorian writing on mad geniuses, although crazed biologists abounded in late-Victorian fiction. Psychologists and physicians like Lombroso and Sully focused on insanity among literary authors, musicians, painters, and the occasional mathematician, rarely scrutinizing geniuses from their own disciplines. For instance, Sully included only one biologist, Georges Cuvier, in his influential 1885 article "Genius and Insanity."[40] Galton, whose own talents were so pronounced, took a kinder view of scientific genius than the other writers represented. Alternatively, some scientists may have actually coveted the cachet of mystery and power surrounding the Romantic genius, even once this aura became tarnished through association with evolutionary undesirables.

How scientists discussed genius changed significantly over the course of the nineteenth century, due to shifting scientific paradigms as much as public sentiment. Early-nineteenth-century authors related genius to monomania, a diagnosis defined by French psychiatrist Jean Esquirol around 1810 that remained popular for several decades. According to Jan Goldstein, monomania was an "idée fixe, a single pathological preoccupation in an otherwise sound mind" that could prompt "overweening ambition."[41] By contrast, late-nineteenth-century authors like Nisbet, Galton, Lombroso, and Max Nordau associated genius with evolutionary regression, asymmetrical brain development, and medical conditions such as hysteria and epilepsy.

French psychiatrist Jacques Moreau's *Morbid Psychology* (1859) became the most influential early treatise on genius and insanity. Moreau, who is today best known for his experiments with marijuana, trained at the Salpêtrière under Philippe Pinel and practiced medicine at the Bicêtre Hospital in Paris. He almost certainly served as the model for Wells's villainous biologist of the same name. Moreau argued in *Morbid Psychology* that "genius was essentially a 'névrose' or nervous affliction similar to idiocy," based on a review of around 180 alleged geniuses.[42] He sought to replace the exalted Romantic view of genius with a vision of great thinkers as diseased victims of biological determinism. Moreau accordingly suggested that geniuses create due to instinct or compulsion, rather than divine inspiration:

> Contrary to what one observes in men of average intelligence, the work of superior men is entirely spontaneous, and in some ways as involuntary as possible. It is the result of impulse and an instinctive need, and of an intellectual appetite that makes itself felt, no one knows why … it has been said, and with reason, that no one is less free in his work, to choose the time of his work in particular, than the men of whom we speak.[43]

Moreau's comments stigmatized the genius as an unfortunate, if occasionally useful, biological anomaly. He further suggested that "the idiot, the hysteric, the epileptic, the madman, as well as the genius, are… branches growing from the same tree"[44] (see Figure 6).

Moreau greatly influenced later continental literature on genius, particularly Lombroso's *The Man of Genius* and Max Nordau's *Degeneration*

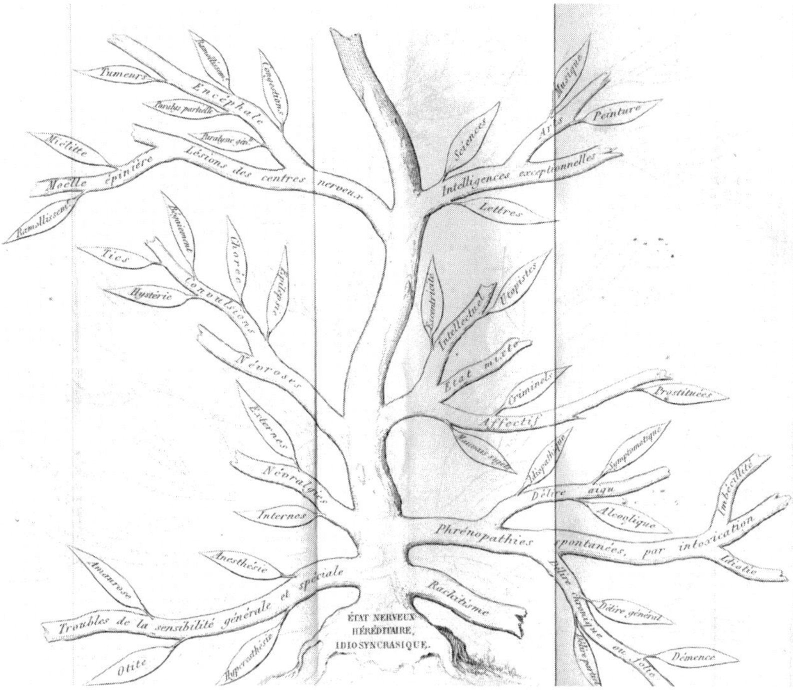


Figure 6 Jacques Moreau's "Tree of Nervosity."

(1892, English translation 1895). Though these works were largely derivative, both have received far more critical attention than Moreau's more original *Morbid Psychology*. This neglect may stem from the relative scarcity of the latter volume, which was never translated or reissued, or from the sensational tone adopted by Moreau's successors. "The man of genius is a monster," claimed Lombroso, "but even monsters follow well-defined … laws."[45] Lombroso contended that "Genius is a true degenerative psychosis belonging to the group of moral insanity."[46] Moral insanity was a common nineteenth-century term for "madness consisting in morbid perversion of the natural feelings, affections … and natural impulses," in which the sufferer remains rational and experiences no hallucinations.[47]

Lombroso solidified the identification between genius and criminality that became a popular theme among late-Victorian scientists and novelists. As with all criminal types he analyzed, Lombroso identified physical signs of genius, especially "elevation of the forehead, notable development of the nose and of the head, [and] great vivacity of the eyes."[48] These stigmata reflected the craniological principle that intelligent people possess developed frontal lobes, pronounced foreheads, and bulging eyes. Lombrosian stigmata quickly became standard features of mad scientist literature, beginning with Stevenson's Hyde, who bears on his body "an imprint of deformity and decay," and continuing with Wells's large-headed, large-eyed mad scientists and extraterrestrials.[49]

Nordau was Lombroso's intellectual disciple who took aim at artistic and poetic geniuses, claiming that they manifested the same degenerate mental and somatic traits as hysterics, criminals, prostitutes, anarchists, and other undesirables. Nordau stridently criticized symbolist and decadent writers like Joris-Karl Huysmans and Henrik Ibsen, not to mention avant-garde musicians like Richard Wagner, whom he classified as "highly gifted degenerates."[50] In one particularly hysterical passage, he asked readers to imagine degenerate protagonists of Ibsen's and Huysmans's works "in competition with men who rise early, and are not weary before sunset, who have clear heads, solid stomachs, and hard muscles." Nordau concluded that "degenerates must succumb," damning the artistic genius to speedy evolutionary extinction.[51] Nordau evidently took Quételet's idealization of the "average man" to an imaginative extreme, envisioning the ordinary Austrian peasant laborer triumphing over the effete, artistic intellectual.

The first important English writer on genius was Galton, Charles Darwin's cousin and an acknowledged genius in his own right, if posthumous IQ scores are to be believed. In 1917, American intelligence-testing pioneer Lewis Terman estimated Galton's IQ at over 200, though one must certainly question the means by which he arrived at this figure.[52] Galton's *Hereditary Genius* advanced the then-controversial notion that genius could be passed down to offspring. Though Galton saw genius as beneficial to humanity, he still worried that "genius … is perilously near to voices heard by the insane, to their delirious tendencies, or to their monomanias."[53] Like Nordau and Lombroso, Galton thought extraordinary men were relatively unlikely to reproduce, blaming the "shy, odd manners, often met with in young persons of genius."[54] Highly intelligent women likewise reproduced infrequently, Galton alleged, since they tended to be "of a dogmatic and self-asserting type, and therefore unattractive to men."[55]

These various writers turned to Lamarck's first and second laws to explain why genius was associated with madness, degeneracy, and infertility. Daniel Pick has emphasized that certain aspects of Lamarckian thought remained scientifically current until the early twentieth century.

Despite the crucial non-Lamarckian concept of natural selection, Darwin himself and many of his followers maintained some Lamarckian assumptions, including (to a limited extent) the heritability of acquired characteristics.[56] In the 1890s, neo-Lamarckian evolutionary theories peaked in popularity as some intellectuals became disenchanted with the materialist implications of Darwinian natural selection, which seemed to permit no human control of evolutionary processes. Neo-Lamarckians hoped to positively influence the future of humanity by acquiring desirable traits and passing them on to their offspring, allowing for more rapid and volitionally directed evolutionary progress than Darwinian natural selection.[57]

Writers linking genius and insanity reasoned according to these Lamarckian principles. For instance, Lombroso suggested that mankind evolved large brains by eliminating unnecessary organs present in lower animals: "Reptiles have more ribs than we have; quadrupeds and apes possess more muscles than we do, and an entire organ, the tail, which we lack. It has been in losing these advantages that we have gained our intellectual superiority."[58] Sully and Lombroso predicted that such lopsided evolution would ultimately result in geniuses who were "puny and ill-formed men" exhibiting "excessive asymmetry of face and head… smallness or disproportion of the body, left-handedness, [and] stammering" among other marks of degeneracy.[59]

Lamarckian principles even surfaced in Victorian educational psychology, which promoted mental balance by avoiding intellectual over-specialization. Alexander Bain recommended exercising the brain to develop "the plastic power of the mind," but cautioned that overusing any particular intellectual faculty could compromise "the whole man."[60] Similarly, Charles Édouard Brown-Séquard advocated ambidextrous education for all schoolchildren in the 1870s in order to prevent bilateral brain hemisphere asymmetry (then thought to cause mental instability).[61]

Neo-Lamarckian thinkers further suggested that hypertrophy of the intellect led to atrophy of the emotions and consequent insanity. Lombroso stated: "Just as giants pay a heavy ransom for their stature in sterility and relative muscular and mental weakness, the giants of thought expiate their intellectual force in degeneration and psychoses."[62] One possible conclusion of rapid Lamarckian brain evolution, then, was a species of morally insane beings boasting enormous cerebrums and minuscule bodies, a nightmarish vision dramatized in Wells's fiction.

Previous scholarship suggests that Wells abandoned Lamarckism around 1895 in favor of German biologist August Weismann's neo-Darwinian, proto-genetic germ plasm theory of inheritance.[63] Weismann argued in 1886 that hereditary information is passed down to offspring via so-called "germ cells" that are distinct from other bodily cells. Moreover, by cutting off the tails of several generations of mice and observing that their offspring still had tails, Weismann apparently demonstrated that acquired traits are not heritable.[64] In early evolutionary writings such as "Incidental Thoughts on a Bald Head" (1895), Wells apparently agreed with Weismann: "Professor Weissman [*sic*] has at least convinced scientific people… that the characters acquired by a parent are rarely, if ever, transmitted to its offspring."[65] But even if one views this indirect statement as indicative of Wells's own beliefs, I suspect that Wells abandoned Lamarckism slowly and reluctantly, rather than all at once. In "Human Evolution, an Artificial Process" (1896) for instance, Wells struck an uneasy compromise between Lamarckism and Weismannism. In this article, Wells acknowledged "Professor Weismann's destructive criticisms of the evidence for the inheritance of acquired characters" but suggested that "artificial" evolution, in the form of education and cultural traditions, could ensure human progress.[66]

Wells's fictional output between 1895 and 1901 likewise bears witness to his continued interest in Lamarckian evolutionary thought. In *The War of the Worlds* and *First Men in the Moon*, massive alien cerebrums evolve at the expense of dwindling, underused bodies, following Lamarck's first law. Wells, however, broke with progressivist neo-Lamarckian thought that emphasized purposeful acquisition of new characteristics as "an intelligence-driven process."[67] Instead, Wells presented a disturbing vision of the dystopian possibilities of Lamarckian brain evolution run amok, as did most late-Victorian writers discussing the pathology of genius.

Lamarckism likewise surfaced in Victorian discussions about female geniuses, who were conspicuously underrepresented in Wells's fiction and in Victorian culture more generally. Nineteenth-century scientists agreed that "genius of the highest order is practically limited to the male sex," only occasionally making exceptions for female novelists like Charlotte Brontë, George Eliot, and George Sand (one suspects the latter two benefited from their willingness to take male pseudonyms).[68] French social psychologist Gustave Le Bon went so far as to state that highly intelligent women "are as exceptional as the birth of any monstrosity, as, for example, of a gorilla with two heads."[69] The few women recognized as geniuses were thought to be a danger to their species. Henry Maudsley, a prominent English psychologist, famously contended that women developed their intelligence at the expense of their reproductive organs, thereby threatening the future of the race.[70] This argument, like the discourses about large-brained male geniuses, stemmed from the Lamarckian idea that overuse of any organ compromised other bodily functions.

The foregoing discussion suggests how racial, national, and gender stereotypes undergirded theories about genius and lent them a measure of cultural authority. While the genius described by Victorians was definitively male, his masculinity was undermined by the suggestion of hysterical effeminacy and his refusal of heterosexual procreation. Moreover, his social and political loyalties came under scrutiny. Lombroso declared, "in men of genius, the love of family and country is either absent or less strong than in other men."[71] Indeed, the fictional trope of the mad scientist traded upon the fear of the scientific intellectual as a transnational, cosmopolitan figure without loyalty to God or country. It is no accident that this rootless being resembled the stereotype of the "wandering Jew." Mad scientists in Victorian fiction often boasted names and physical features stereotypically associated with German-speaking Eastern European Jews, reflecting English fears of German invasion and anti-Semitic backlash against high levels of Jewish immigration to England during the nineteenth century.[72] These cultural stereotypes, buttressed by scientific authority, crystallized in late-nineteenth-century fictional portrayals of mad scientists.

Wells wrote his early scientific romances during the decade when interest in mad geniuses peaked in the popular press. His fictions faithfully adhere to the neo-Lamarckian evolutionary theory expressed in these discourses, but add a fascinating new twist by morphing the mad scientists of *The Island of Doctor Moreau* and *The Invisible Man* into the top-heavy extraterrestrials of *The War of the Worlds* and *The First Men in the Moon.* The transition is seamless, striking, and utterly novel, reminding us just how threatening superior intelligence seemed even at a time when it was increasingly necessary to professional success and scientific progress.

## MAD SCIENTISTS AND GIANT BRAINS FROM OUTER SPACE

One might reasonably inquire which late-Victorian writings on genius influenced Wells's fiction. Though Wells read Lombroso and Nordau, I suspect that he was particularly receptive to Galton, with whom he corresponded on occasion, and to Nisbet, whom he met in the late 1890s.[73] In 1899, Wells reviewed Nisbet's volume *The Human Machine* for *The Academy* (1869–1916). In his review, Wells lamented that *The Insanity of Genius* received insufficient recognition: "Nisbet's former work, on *The Insanity of Genius*, temperately and soundly argued, was comparatively speaking a failure. Dr. Nordau's bawling version of the same thesis, coarsely seasoned with gross personalities, sauced with a dressing of sexual incontinence … attained a vast success."[74] Wells's admiration notwithstanding, *The Insanity of Genius* is now forgotten, while *Degeneration* is still in print over one hundred years later. Critical neglect of Nisbet's work and of his relationship with Wells have occluded important links between *The Insanity of Genius* and *The Island of Doctor Moreau*. Since Nisbet quoted extensively from Jacques Moreau's *Morbid Psychology*, this real-life Doctor Moreau was very likely the prototype for Wells's fictional namesake.[75]

Galton's work on genius may have had an even more pronounced influence on Wells's early fiction, however. Along with Huxley and Herbert Spencer, Galton was one of the three "leading 'men of science'" whose thinking surfaced most frequently in Wells's scientific romances (even if Wells sometimes disagreed with their ideas).[76] For instance, Wells invokes Galton's eugenics when the Morlocks selectively breed the Eloi in *The Time Machine*. In his later romance, *A Modern Utopia* (1905), Wells borrowed Galton's idea of using fingerprinting for identification, and launched a sustained critique of his eugenic principles.[77]

While Lombroso, Nordau, and others described geniuses as monstrous deviants, Galton took a more sympathetic attitude toward gifted individuals. Far from seeing such persons as morally deficient and socially isolated, Galton stated that "great men have unusually high moral natures, and are affectionate and reverential, inasmuch as mere brain without heart is insufficient to achieve eminence."[78] Moreover, while Galton acknowledged that geniuses tend to reproduce less successfully than more mediocre types, he disagreed with the prevailing notion that "men of genius are unhealthy, puny beings – all brain and no muscle."[79] To prove his point, Galton asserted that senior wranglers at Cambridge "have frequently been the first oarsmen of their years."[80]

Judging from the scientist villains in his fiction, Wells seems to have held the widespread view that genius usually accompanies some sort of physical or psychological deficiency. Yet Wells shared with Galton a sense of the inherent value of genius. In *English Men of Science*, Galton advocated "the establishment of a sort of scientific priesthood throughout the kingdom, whose high duties would have reference to the health and well-being of the nation."[81] In a similar vein, Wells's first full-length sociological work, *Anticipations* (1901), predicted the eventual rise of a class of scientific experts who would provide benevolent leadership to the "world state" of the future. Just as Galton connected high intelligence and ethical conduct, Wells emphasized that this capable expert class "will certainly be a moral people," since "they will have in their positive science a common ground for understanding the real pride of life, the real reason for the incidental nastiness of vice."[82] Appropriately, both Galton and Wells promoted educational reforms that would encourage greater scientific literacy throughout the British population, and induce more gifted students to pursue scientific careers.[83]

Wells combined his appreciation for scientific genius with compassion for the psychological, social, and physical debilities supposedly afflicting the most highly talented individuals. While Wells predicted that scientific experts of the future would be eminently sane and rational, he also expressed concern for a few beings at the summit of intellectual achievement, those "precociously brilliant creatures – creatures whose brilliance is too often the hectic indication of a constitutional unsoundness of mind."[84] Wells's sympathy for such unfortunate prodigies is palpable throughout his early scientific romances. His primary intervention in late-Victorian discourses on genius and insanity, I will suggest, was to lend mad geniuses an element of humanity denied them in many other discussions of the subject. This remains true even in later narratives like *The War of the Worlds* and *The First Men in the Moon*, once the geniuses in question have literally ceased to be human. These aliens differ from human geniuses in degree more than kind, since they have developed their large brains by a process analogous to the human evolutionary trajectory set forth in "The Man of the Year Million."

There are several possible reasons why Wells displayed compassion for the insane geniuses vilified in scientific discussions of his day. Perhaps the author's personal experiences as a misunderstood intellectual, combined with his reading of Galton, gave him a measure of sympathy for such individuals. Indeed, many of Wells's early works depict the unjust social prejudice against geniuses (particularly scientific prodigies) fomenting during the late-Victorian period. Alternatively, Wells's relatively sympathetic portrayal of mad geniuses may relate to the didactic mission of his early scientific romances. One goal of these early fictions, I argue, was to warn late-nineteenth-century readers about the disturbing evolutionary consequences of rapid Lamarckian brain development. A future in which human beings resembled "tadpoles" with enormous heads, tiny bodies, and "great unemotional intelligences," was certainly undesirable in many ways.[85] To the extent that Wells believed in Lamarckian or "artificial" evolutionary processes, it made sense for him to recommend a volitionally directed course of evolution for human beings, or to warn against following the wrong evolutionary path. But if Wells's readers could not see themselves in his large-headed mad scientists and alien invaders, the lessons of his fiction would be lost on them. In order to heed Wells's evolutionary warnings, readers would need to have at least some measure of identification with these strange and sometimes monstrous beings.

Beginning in his first published fiction, the short story "The Chronic Argonauts" (1888), Wells depicted a series of mad scientist characters whose behavior ranges from eccentric to malevolent. Despite the disturbing, even monstrous attributes of these geniuses, Wells's narratives stop short of condemning these characters outright. Instead, each of these fictions allows for a certain moral ambiguity in which the genius may be seen alternately as malign or as a misunderstood victim of circumstance or social prejudice. This moral ambiguity pervades Wells's portrayal of Dr. Moses Nebogipfel, Ph.D., F.R.S., the eccentric protagonist of "The Chronic Argonauts."[86] This short story, written when Wells was only twenty-one, was a very early draft of what later became *The Time Machine*. But the tale also prefigures *The Invisible Man* in important ways, particularly in its portrayal of a scientific outcast.[87] At the story's opening, Nebogipfel retires to an isolated village in Wales to pursue his arcane research in time travel. He succeeds in building a time machine, but his odd appearance and strange, noisy experiments provoke the suspicions of the villagers. Incensed by the mysterious death of the town hunchback, the villagers pursue Nebogipfel with torches and threaten to burn him for witchcraft. The cornered doctor finally escapes this angry mob by heading off into the distant future on his time machine.

The story's opening passages foreground the doctor's strange appearance, emphasizing the Lombrosian physical stigmata popularly associated with genius:

> He was a small-bodied, sallow faced little man … [h]is aquiline nose, thin lips, high cheek ridges, and pointed chin, were all small and mutually well-proportioned; but the bones and muscles of his face were rendered excessively prominent and distinct by his extreme leanness. The same cause contributed to the sunken appearance of the large eager-looking grey eyes, that gazed forth from under his phenomenally wide and high forehead. It was this latter feature that most powerfully attracted the attention of an observer. It seemed to be great beyond all preconceived ratio to the rest of his countenance … below it his eyes glowed like lights in some cave at a cliff's foot. It so overpowered and suppressed the rest of his face as to give an *unhuman* appearance.[88]

This lengthy passage repeatedly draws attention to the doctor's huge eyes and bulging forehead, not to mention his disproportionately small body and massive cranium. These features are so striking that the doctor appears "unhuman," suggesting the beginnings of the metamorphosis from mad scientist to extraterrestrial that is fully realized in *The War of the Worlds* and *The First Men in the Moon*. Alternatively, the doctor may simply be more evolved than the average human, having somehow attained a bodily form closer to that of the beings described in "The Man of the Year Million."

Nebogipfel entertains the latter hypothesis, describing himself as "a man born out of my time." In a surprisingly plaintive passage, the doctor tells a local clergyman about the sufferings he has undergone due to his unusually high intelligence:

> In short, Mr. Cook, I discovered that I was one of those superior Cagots called a genius – a man born out of my time – a man thinking thoughts of a wiser age, doing things and believing things that men now *cannot* understand, and that in the years ordained to me there was nothing but silence and suffering for my soul – unbroken solitude, man's bitterest pain. I knew I was an Anachronic Man; my age was still to come.[89]

The purpose of the time machine he is building, Nebogipfel explains, is to help him "join my generation, to journey through the ages till my time has come."[90] Exactly how or why Nebogipfel managed to be born into the wrong time is left unexplained.

Here and elsewhere, Wells's portrayal of Nebogipfel draws attention to the isolation and emotional pain allegedly suffered by geniuses. Earlier in the story, Nebogipfel compares himself to the "Ugly Duckling" of Hans Christian Andersen's fairy tale: "a wonderful story – a story that has ever been full of tears and heart swelling hopes for me, since it first came to me in my lonely boyhood." The story leads him to dream of "encountering the sympathy I knew was my profoundest need."[91] But Nebogipfel's immense intellect, and the social and physical defects that come with it, shut him off from the human fellowship he craves.

Like his first protagonist, Wells was born into an environment that was uncongenial to his prodigious mental gifts. In his 1934 *Experiment in Autobiography*, Wells describes how his lower-middle-class parents discouraged him from following his intellectual ambitions, urging him to pursue a steady trade in retail or pharmacy instead.[92] "The atmosphere of my home and early upbringing was not … highly educative," Wells wrote with characteristic understatement, referring to the "quite dreadful conditions" in which he spent his first years.[93] Wells transcended this background, however, by receiving a scholarship to attend the Normal School of Science, and later by writing bestselling fiction. It is tempting to compare Nebogipfel's experiences as a man born into the wrong time with Wells's own circumstances. As a man born into the "wrong" socioeconomic environment, Wells found it exceedingly difficult to nurture his intellectual gifts. These biographical considerations suggest why Wells's depiction of the plight of the mad genius was so moving in his first work of fiction, when he was still chronologically close to these early experiences.

Wells's subsequent fictions about mad geniuses, *The Island of Doctor Moreau* and *The Invisible Man*, are markedly less sympathetic toward their scientist villains. Whereas Nebogipfel appears misunderstood and socially awkward, rather than malevolent, Moreau and Griffin undertake their unusual experiments out of a combination of intellectual curiosity and desire for mastery.[94] Careless of the pain they inflict on others – or even sadistically delighting in that pain – Moreau and Griffin seemingly merit moral condemnation, not pity. But even so, Wells repeatedly emphasizes these scientists' vulnerability to social prejudice, to the scientific ignorance of the public, and to the varieties of mental illness thought to afflict geniuses. Since both characters die in the course of their narratives, it is even possible to see them as martyrs for the cause of scientific progress. By portraying scientific villains in this relatively non-judgmental fashion, Wells lent a human dimension to late-Victorian discussions of genius and insanity, which were apt to depict such talented individuals as monsters. He also made it possible for readers to occasionally identify with Moreau and Griffin, thereby recognizing their own stake in the evolutionary warnings implicit in these novels.

In *The Island of Doctor Moreau* and *The Invisible Man*, Wells focused less on the physical stigmata of genius foregrounded in "The Chronic Argonauts," while still hinting at the Lamarckian evolutionary trajectory outlined in "The Man of the Year Million" and *The Science of Life*. Whereas Nebogipfel's strange appearance turns villagers "rigid with horror," neither Moreau nor Griffin is strikingly deformed.[95] Moreau possess a "fine forehead" bespeaking large frontal lobes and superior intellectual gifts, while Griffin's albinism ensures the success of his experiments in invisibility.[96] Aside from these unusual features, however, both characters' robust physiques belie received opinions about weakly, ill-proportioned geniuses. Moreau has a "magnificent build," while Griffin exhibits unusual physical strength, bringing to mind Galton's refutation of the puny genius stereotype.[97] Perhaps Wells scaled back the physical abnormalities of Moreau and Griffin in order to make his scientific romances of the mid 1890s more realistic in comparison to his juvenile effort, which he judged to be "overly elaborate" and "rococo."[98]

For the most part, Wells's depictions of Moreau and Griffin focus on moral and psychological abnormalities attendant on genius, rather than physical ones. These psychological abnormalities must be inferred from behavior. As narrator Edward Prendick explains, Moreau is a "prominent and masterful physiologist" forced to leave England due to public outcry over his "wantonly cruel" experiments involving animal vivisection.[99] Moreau settles on remote Noble's Island near the Galapagos and there grafts the bodies of various kinds of animals together in order to create "Beast People" with near-human intelligence. These monsters speak broken English and worship their creator. Like Victor Frankenstein, Moreau is unmarried, single-mindedly devoted to research, and shamefully neglectful of his creations, one of whom (a female puma) ultimately kills him.[100]

Wells implies that Moreau's intellect has evolved far enough to compromise his emotional sensitivity. His monomaniacal quest "to find out the extreme limit of plasticity in a living shape" outweighs all considerations of suffering it might cause. When Prendick asks how the doctor can bear the screams of his vivisected animals, Moreau explains: "The thing before you is no longer an animal, a fellow-creature, but a problem. Sympathetic pain – all I know of it I remember as a thing I used to suffer from years ago."[101] Moreau's appalling indifference suggests the "complete absence of moral sense and of sympathy" that Lombroso associated with moral insanity.[102]

Early drafts of *The Island of Doctor Moreau* indicate that Wells imagined the doctor's amorality and single-minded devotion to experimentation to result from mental illness, probably some variety of monomania. In one rough draft, Moreau's assistant, Montgomery, excuses his employer's scientific preoccupation on the grounds of mental instability: "This research is only a sane kind of mania. It's irresistible. He's driven to make these things, can't help it any more than an avalanche … can help smashing a tourist."[103] This passage exemplifies the late-Victorian biological determinist view of mental illness as hereditary and therefore untreatable, a position championed by scientific luminaries such as Moreau and Maudsley. Seen from the grim perspective of evolutionary neurology, then, Doctor Moreau's overdeveloped rationality is the monstrous presence on the island, not the grafted hybrids he creates.

But despite his shocking emotional detachment, Moreau never becomes a one-dimensional caricature, unlike the evil geniuses of some lesser Victorian fiction.[104] Readers gain insight into Moreau's motivations by learning about the persecution he has endured in his native England. Prendick explains how, some years prior to the novel's events, Moreau "had published some very astonishing facts in connection with the transfusion of blood, and, in addition, was known to be doing valuable work on morbid growths." Moreau's illustrious career suddenly comes to a close, however, when he is "howled out of the country" following a journalistic exposé of his unprincipled use of vivisection. Prendick implies that Moreau was framed by writers and editors in search of a sensational story: "A journalist obtained access to his laboratory … with the deliberate intention of making sensational exposures; and by the help of a shocking accident – if it was an accident – his gruesome pamphlet [on vivisection] became notorious."[105]

Although Wells supported the responsible use of vivisection, his attitude toward Moreau's cruel experiments seems ambivalent at best.[106] Nonetheless, we are certainly meant to sympathize with the scientist at this juncture, as Laura Otis suggests: "In Wells's representation … readers cannot tell whether Moreau's London experiments were truly sadistic. It was the anti-vivisectionists who acted like cruel, baying hounds."[107] In Prendick's account, the British public's ignorance about science and appetite for lascivious stories condemn Moreau to a Crusoe-like existence on a deserted island. Moreau's precarious mental state following these events underscores the cruelty of his fate. Perhaps the sadistic, monomaniacal Moreau Prendick sees is very different from the Moreau who lived and worked in England, since it is unclear how much of his mental dysfunction results from his exile.

Also at issue here is the British government's treatment of scientists, whose research methods were severely limited by the 1876 Anti-Vivisection Act. In *Anticipations*, Wells complained that government officials typically ignored the advice of scientific experts, except when trotting them out as a kind of spectacle: "The man of special equipment is treated always as if he were some sort of curious performing animal," Wells lamented.[108] He continued: "The modern democracy, or democratic quasi-monarchy, conducts its affairs as though there was no such thing as special knowledge or practical education. The utmost recognition it affords to the man who has taken the pains to know, and specifically to do, is occasionally to consult him upon specific points and override his counsels in its ampler wisdom."[109] Moreau certainly experiences the indignity of becoming a "spectacle" and being treated like an animal, hunted and hounded out of his own country. His plight symbolizes that of *fin-de-siècle* scientists more generally in that his contributions are misunderstood and undervalued, while his research methods are hampered by the intervention of non-experts.

Moreau can be seen, then, as a victim of social prejudice against scientific experts and individuals of exceptionally high intelligence. In fact, Wells wrote *The Island of Doctor Moreau* with another brilliant victim of social prejudice in mind. In 1924, Wells reflected that the 1895 trial and conviction of Oscar Wilde had formed an important backdrop to his tale: "There was a scandalous trial about that time, the graceless and pitiful downfall of a man of genius, and this story was the response of an imaginative mind."[110] While both Moreau and Wilde have broken the law – vivisection was illegal after 1876, unless one possessed the proper licenses – they are reviled mainly because they have violated social taboos of one sort or other.[111] Wilde's homosexuality flouted the social customs of his day, as does Moreau's eccentric behavior. The scientist's unemotional nature, refusal to marry, and monomaniacal interest in his research may have contributed to his exile as much as any specific legal transgression.

In *The Invisible Man*, the fictional trope of the mad scientist is perhaps more fully expressed than anywhere else in Wells's oeuvre. Griffin is a megalomaniac bent on world domination, who commits robbery, assault, and murder in order to further his quest for power. He begins by stealing money from his father to fund his research into invisibility, which is conducted with an eye to personal gain rather than social utility. Once invisible, he feels "a wild impulse to jest, to startle people, to clap men on the back, fling people's hats astray, and generally revel in my extraordinary advantage."[112] After he has tired of such puerile amusements, Griffin dreams of starting "A Reign of Terror" in which he will "terrify and dominate" a rural British town by going on a killing spree.[113] Griffin certainly demonstrates the moral insanity described by Nisbet, Nordau, and others embroiled in *fin-de-siècle* discussions of mad geniuses. To this diagnosis one might possibly add monomania – or even what we would now call psychosis – due to Griffin's extravagant delusions of grandeur in the novel's concluding chapters.[114]

Griffin does not bear the Lombrosian physical stigmata of rapid brain evolution. Nonetheless, his physicality persistently reasserts itself throughout the tale, in a manner that alternately highlights his potential monstrosity and underscores his vulnerability. Griffin's deformities (both real and imagined) remind us of his status as social outcast and suggest a possible reason for his misdirected anger and malignity. His albinism plays little overt role in the text, other than making his experiment in invisibility more plausible. But the physical strangeness of Griffin's disguise causes some Iping villagers to guess at his true condition. Fearenside remarks, "That marn's a piebald … black here and white there – in patches. And he's ashamed of it. He's a kind of half-breed, and the colour's come off patchy instead of mixing. I've heard of such things before."[115] Although Fearenside does not say where he has "heard of such things," albinos and others with pigmentation disorders featured prominently in late-Victorian traveling freak shows.[116] This fact suggests one powerful motivation behind Griffin's quest for invisibility, and implicitly links geniuses to other, more traditional types of "monsters."

But Griffin's invisibility comes with a high physical cost, as he explains to his former classmate, Dr. Kemp: "I had no shelter, no covering. To get clothing was to forgo all my advantage … I was fasting; for to eat, to fill myself with unassimilated matter, would be to become grotesquely visible again."[117] These privations become most serious when Griffin flees from police and angry townspeople. The scientist's protracted physical struggle with cold and hunger underscore his humanity, even as he is pursued "like a rabbit hunted out of a woodpile."[118] His trials end when he is savagely beaten to death by an angry mob. As Griffin's corpse slowly becomes visible, his pursuers see "naked and pitiful on the ground, the bruised and broken body of a young man about thirty."[119] In death, Griffin's perishable humanity reasserts itself to those who had dismissed him as a freak, a monster, or a supernatural menace.

While emphasizing Griffin's vulnerability, *The Invisible Man* also subtly suggests the threatening aspects of Griffin's genius, by hinting at his rapid Lamarckian brain development and concomitant psychological and physical degeneration. Anatomically, Griffin is a human being of relatively normal proportions, whose head size is limited because the skull must be small enough to fit through the female birth canal. But Griffin's uncanny disguise foreshadows the large-headed, bug-eyed aliens and future humans that could result from unchecked Lamarckian brain development. His costume when invisible includes layers of bandages covering his head, not to mention huge blue goggles, causing the townspeople of Iping to liken him to "a lobster" or "a divin' helmet."[120] Like the more evolved alien beings of Wells's later scientific romances, Griffin sometimes fills passersby with a nameless sense of dread; on more than one occasion, his "goggling spectacles and ghastly bandaged face" frighten laborers and children returning home at dusk.[121]

After observing this peculiar new guest at her hotel, the landlady of the Coach and Horses has a dream in which the striking features of this disguise are especially pronounced: "The stranger was undoubtedly an unusually strange sort of stranger, and she was by no means assured about him in her own mind. In the middle of the night she woke up dreaming of huge white heads like turnips, that came trailing after her at the end of interminable necks, and with vast black eyes."[122] Mrs. Hall's nightmare vision of the Invisible Man resembles nothing so much as the illustration accompanying *Punch*'s version of "The Man of the Year Million." Even that seemingly incongruous detail – "interminable necks" – evokes the detachment between brain and body (or the absence of body) that Wells foresaw as part of our evolutionary future. The landlady's troubled dream suggests Griffin as the "missing link" between the humans of today and the more highly evolved human or alien beings whose brains are not limited by bone structure.

In his next works, *The War of the Worlds* and *The First Men in the Moon*, Wells did away with the limitations of the human form altogether in an effort to more fully explore the possibilities of Lamarckian brain development. These novels literally *alienate* geniuses by morphing them into grotesque extraterrestrials. The language of alienation permeating late-Victorian discourse about genius set the stage for this characterological transformation. Sully, for example, declared that "the peculiar leanings and aspirations" of the man of genius "stamp him as an alien."[123]

By turning his extraterrestrial geniuses into frightening physical grotesques, Wells ran the risk of alienating them from our sympathy. But Wells left room for readers to identify with Martians and lunar beings (also called Selenites) by endowing them with a number of positive qualities. On the one hand, Wells's Martians and Selenites seemingly usurp the stereotyped role of the mad scientist with their hypertrophied intellects, atrophied emotional sympathies, and unconcealed impulse toward world domination. They are all the more terrifying for being vastly more effective than Wells's previous scientist villains, as Bernard Bergonzi suggests: "The Martians establish the 'Reign of Terror' that the Invisible Man could only dream of – and on an infinitely larger scale."[124]

On the other hand, both the Martians and Selenites display characteristics that readers might pity or admire, so that neither group fully or comfortably inhabits the mad genius stereotype. For instance, the Martians' cooperation contrasts favorably with the brutally individualistic behavior of their human victims. The Selenites, meanwhile, display many of the positive attributes of the scientific "expert class" Wells described in *Anticipations*, which was published during the same year as *The First Men in the Moon*.[125] For all of their failings, both Martian and Selenite societies possess characteristics human beings might wish to emulate, alongside evolutionary flaws we would do well to avoid. Following Lamarckian logic, sagacious readers of these novels could intuit which traits, behaviors, and attitudes to encourage or discourage in themselves and their offspring, so as to forestall the disastrous evolutionary futures Wells predicts.

The invading Martians of Wells's *The War of the Worlds*, with their advanced civilization and enormous brains, present a dystopian vision of what might happen if *all* humans evolved into geniuses. One obvious consequence of this development would be widespread moral insanity. To their human victims, the Martians' amorality becomes evident soon after they arrive on Earth and mercilessly slaughter the human welcoming party sent to greet them. They later consume the desperate survivors who remain after they destroy London with terrifying heat rays and toxic gas. In a short time, the Martians overcome all military resistance and nearly take over the Earth. These giant brains from outer space seemingly epitomize the mad scientist trope, in which the conscienceless, cosmopolitan scientist achieves world domination and wreaks havoc on human subjects.

To be sure, Wells presents his invading Martians first and foremost as warriors rather than scientists. But their technological advancements testify to their skill as inventors, while their "intellects vast and cool and unsympathetic" suggest their conformity to the mad scientist trope Wells explored in his earlier scientific romances. Moreover, the Martians are compared to scientists in the novel's opening paragraph, where they are said to study the human race "perhaps almost as narrowly as a man with a microscope might scrutinize the transient creatures that swarm and multiply in a drop of water."[126] The Martians are also compared to Western imperialists when the narrator likens their treatment of human beings to the English extermination of the Tasmanians.[127] On these grounds, Bergonzi and other critics have viewed the novel as a potent critique of imperialism, as seen through the eyes of the colonized population.[128]

But the novel also contains another, perhaps more urgent warning about a possible evolutionary trajectory for human beings. Upon encountering his first Martian, the narrator relates:

> Two large dark-coloured eyes were regarding me steadfastly. The mass that framed them, the head of the thing, was rounded, and had, one might say, a face. There was a mouth under the eyes, the lipless brim of which quivered and panted, and dropped saliva. The whole creature heaved and pulsated convulsively.

A lank tentacular appendage gripped the edge of the cylinder, another swayed in the air.[129]

Peter Kemp observes that this view of the Martians emphasizes the slobbering mouth, drawing attention to the aliens' prodigious appetites.[130] But the Lombrosian stigmata of genius are also evident in this portrayal, especially the large eyes and apparently massive head. As in Wells's earlier description of Dr. Nebogipfel, these stigmata are so pronounced that the resulting form only partially resembles a human face.

How did the Martians attain this grotesque shape? Rather than leaving this question open to speculation, the narrator playfully alludes to "The Man of the Year Million," suggesting that Martians developed along much the same lines as Wells's future humans:

A certain speculative writer of quasi scientific repute, writing long before the Martian invasion, did forecast for man a final structure not unlike the actual Martian condition. His prophecy, I remember, appeared in November or December, 1893, in a long-defunct publication, the *Pall-Mall Budget*, and I recall a caricature of it in a pre-Martian periodical called *Punch*. He pointed out – writing in a foolish, facetious tone – that the perfection of mechanical appliances must ultimately supersede limbs; the perfection of chemical devices, digestion; Such organs as hair, external nose, teeth, ears and chin were no longer essential parts of the human being … the tendency of natural selection would lie in the direction of their steady diminution through the coming ages. The brain alone remained a cardinal necessity. Only one other part of the body had a strong case for survival, and that was the hand, "teacher and agent of the brain." While the rest of the body dwindled, the hands would grow larger … here in the Martians we have beyond dispute the actual accomplishment of such a suppression of the animal side of the organism by intelligence.[131]

This passage not only alludes to Wells himself ("a certain speculative writer of quasi scientific repute") and his article, but also quotes certain portions of "The Man of the Year Million" almost verbatim. We thus come to understand the Lamarckian evolutionary logic by which the Martians have become "heads – merely heads. Entrails they had none."[132] This passage also suggests why tentacles play a large part in Martian evolution, since they serve a function analogous to the human hand (the "teacher and agent of the brain").[133] *The War of the Worlds* makes abundantly clear that "the Martians may be descended from beings not unlike ourselves," and that their frightening intellectual and physical development possibly foreshadows our own.[134]

In some ways, however, the Martians seem better adapted to life on Earth than human beings, particularly since they are free to develop and expand their intellects without physical hindrance. In "The Man of the Year Million," Wells sidestepped one question he would later address in *The Science of Life*: namely, how could Lamarckian brain evolution proceed, given that the human skull must fit through the female birth canal? Because the Martians reproduce asexually, by "budd[ing] off" from a parent organism, there are no natural limits to the cerebral growth they can achieve.[135] Another advantage of Martian development is their relatively uncomplicated digestive system. The narrator explains how the Martians "did not eat, much less digest. Instead, they took the fresh, living blood of other creatures, and *injected* it into their own veins." This grotesque detail was likely "suggested by the publication of Stoker's *Dracula* in 1897," as Gordon S. Haight notes.[136] By injecting blood instead of eating food, Martians avoid the "tremendous waste of human time and energy occasioned by the eating and digestive process," not to mention the mood swings occasioned by bad digestion.[137]

Notably, both of these advantages are secured by means of evolutionary regression rather than progress. The narrator reminds us that asexual reproduction is typically associated with less complex, "lower" life forms such as plants, bacteria, or "the fresh water polyp." On Earth, at least, asexual reproduction was "certainly the primitive method" as compared to the sexual reproduction of higher animals.[138] Moreover, the Martians' blood-drinking is characteristic of parasites. The future humans of "The Man of the Year Million," are also parasitic, although not bloodthirsty; they nourish themselves by "immersion in a tub of nutritive fluid."[139] According to Wells's friend Edwin Ray Lankester, the author of *Degeneration: A Chapter in Darwinism* (1880), parasitism was a sure sign of evolutionary regression: "Let the parasitic life once be secured, and away go legs, jaws, eyes, and ears; the active, highly gifted crab, insect, or annelid may become a mere sac, absorbing nourishment."[140] The Martians' means of sustenance casts doubt on the evolutionary trajectory they have followed, along with the similar evolutionary scheme outlined in "The Man of the Year Million."

Even the more progressive aspects of Martian evolution have material drawbacks. Following neo-Lamarckian logic, the Martians' accelerated brain evolution weakens their fragile bodies, to the point where they are practically useless. In *The War of the Worlds*, Martian bodies hardly ever become visible, since they are encased in prosthetic devices that provide greater mobility, power, and destructive capacity. These prosthetic devices resemble "a great body of machinery on a tripod stand," a form which in turn suggests the Martian's large heads and elongated limbs.[141] Within these tripod-like machines, the Martians lay waste to England with fearful weapons. Outside of these prostheses, however, the Martians appear "crippled" and move in a "clumsy" fashion, hampered by the Earth's strong gravitational pull.[142] By pursuing a course of perilous intellectual abstraction in which body parts are replaced by mechanical components, the Martians may have further accelerated their development into bodiless heads. According to Colin Manlove, their evolutionary course suggests "that to go forward in technological isolation is to go backward biologically; a race that puts all premium on mind and machine is reduced to the most primitive and frail of physical bodies."[143] Indeed, the Martians' technological advancement has "rendered them … a disabled race, wholly tied to prosthetic supports."[144] The Martians are also consummately vulnerable to disease, since they ultimately succumb, in a neat final paradox, to earthly bacteria to which humans have gradually evolved a resistance.

Such bodily atrophy could lead to moral insanity as well as physical fragility, the narrator reminds us: "Without the body the brain would, of course, become a mere selfish intelligence, without any of the emotional substratum of the human being."[145] But this statement is somewhat misleading, as critics such as W. Warren Wagar and Patrick Parrinder have explained. In some instances, the Martians actually appear morally superior to human beings. "One could readily argue that at least the Martians were well within their rights to exterminate humankind," Wagar explains. "From their perspective, human beings were lower animals occupying choice real estate that the Martians needed for their survival."[146] By contrast, the English extermination of the Tasmanians has no such life-or-death justification. The Martians are also a very cooperative species, who are capable of coordinated military and scientific efforts, and who help one another in times of crisis.[147] They even appear to grieve for their dead and look after their remains, as when a group of fighters retreat to Horsell Common "encumbered with the débris of their smashed companion."[148] The Martians' kindness toward one another contrasts favorably with the brutality of their fleeing human victims, who beat and trample one another in their frantic efforts to escape London. Seen from the perspective of their human food supply, the Martians are of course brutal and terrifying; toward members of their own species, however, the Martians display the utmost consideration.[149]

In general, Wells presents the Martians as a cautionary tale of Lamarckian evolution gone awry. But if readers cannot see themselves in the Martians to some extent, then they might overlook the evolutionary warnings of *The War of the Worlds*. Wells's emphasis on the Martians' kindness and cooperation allows readers to identify with the aliens just enough to grasp our own possible evolutionary future. The Martians' positive qualities might also serve as a reminder not to hastily misjudge earthly geniuses. Here, as in his other scientific romances, Wells was acutely aware of the social prejudice faced by highly intelligent individuals, and subtly worked to undermine the negative stereotypes embodied in Victorian discussions of genius and insanity.

While *The War of the Worlds* fits neatly within the late-Victorian Gothic tradition, Wells's later novel, *The First Men in the Moon*, serves as a biting satire of craniometric principles and the cerebral localization theories upon which they were based. This darkly comic work mocks intellectual over-specialization using the familiar figures of the mad scientist and the top-heavy extraterrestrial. The novel's protagonist, an eccentric chemist named Cavor, is an unmarried and obsessive researcher in the tradition of Frankenstein and Moreau, although he is bumbling rather than malevolent. He becomes the alien visitor from another planet after he invents a gravity-defying substance and journeys to the moon. There, he unexpectedly discovers a complex lunar society of large-brained, insect-like beings called Selenites who live in deep underground caverns, just as Wells predicted in "The Man of the Year Million."[150] Cavor radios messages to Earth conveying his admiration for Selenite culture, but his final broken transmission implies that the Selenites kill him. As in *The War of the Worlds*, the novel consistently parallels human scientific rationality and amorality with that of the aliens. By juxtaposing a human mad scientist with extraterrestrials who exhibit similar mental tendencies, Wells likens monstrous alien life forms to the moral monstrosities of which unscrupulous geniuses are capable.

*The First Men in the Moon* seems very different in tone and intention than Wells's earlier scientific romances, as pointed out by the novel's initial reviewers and by recent critics.[151] While romances like *The Island of Doctor Moreau* and *The War of the Worlds* convey an atmosphere of terror and mystery, *The First Men in the Moon* is a lighthearted satire whose object is sometimes unclear. This narrative ambiguity is most striking in Wells's portrayal of Selenite society. In some respects, the Selenites' social organization seems utopian, as Wagar suggests: "the Moon shelters a rational, orderly, efficient, and peaceful society, spanning the whole cavernous interior of the lunar space … it knows nothing of wars or tribal conflicts."[152] Moreover, the Selenites display many of the positive attributes of the human expert class described in Wells's *Anticipations*, including their cooperation in pursuit of scientific goals, their embrace of applied (as opposed to theoretical) science, and their increased social specialization.[153] As Steven McLean observes, the "global unity" of lunar society may even be seen as a "tentative articulation" of the twentieth-century world state envisioned in Wells's sociological works.[154]

But the Selenites carry their social specialization to ridiculous and sometimes terrifying extremes, as Cavor discovers when he meets a group of embryonic Selenite "machine hands": "Quite recently I came upon a number of young Selenites, confined in jars from which only the forelimbs protruded, who were being compressed to become machine-minders of a special sort ... That wretched looking hand sticking out of its jar seemed to have a sort of limp appeal for lost possibilities; it haunts me still."[155] Here, Cavor seems appropriately critical of the Selenites' social organization, but elsewhere, his attitude toward lunar society resembles Gulliver's infatuation with the Houyhnhnms. Cavor's narrative unreliability makes it all the more difficult to discern Wells's authorial perspective on his curious alien society. It may well be, as McLean suggests, that Wells used Selenite society as "an arena in which he tests his sociological ideas" that he worked out in more didactic form in *Anticipations*.[156] This seems logical, given Wells's proclivity for testing unusual hypotheses within his scientific romances.

Wells also used this novel to test the most extreme implications of Lamarckian brain development that he had explored in his earlier novels. But whereas Wells's earlier romances examined the distress endured by individual geniuses, or explored how highly evolved geniuses of the future might appear to humans of today, *The First Men on the Moon* considers how an entire community of geniuses might interact together. The results range from utopian to frankly disastrous, again suggesting Wells's ambivalence toward the evolutionary course laid out in "The Man of the Year Million." One didactic purpose of this strange novel, then, may be to suggest the *social* consequences of intellectual overdevelopment at the expense of bodily and emotional atrophy.

Selenites come in an incredibly variety of shapes and sizes, as the above example of the machine hand suggests; there are also Selenite herders, guards, policemen, and so forth, all of whom possess a physical form appropriate to their duties. Of greatest interest here are "those beings with big heads to whom the intellectual labors fall, who form a sort of aristocracy in this strange society."[157] As in *The War of the Worlds*, Wells speculates about how such large-headed intellectuals might have evolved to bypass the dilemma of the skull and pelvis described in *The Science of Life*. This time, instead of turning to plants and polyps for a solution, Wells invokes the insect kingdom, as he would do again in his short story "The Empire of the Ants" (1905).[158] Like a vast beehive or anthill, Selenite society boasts queens whose sole function is to give birth to larvae, most of whom are "of the neuter sex."[159] The brains of these larval Selenites can develop to an unlimited extent because they lack skulls. Cavor relates: "The unlimited development of the minds of the intellectual class is rendered possible by the absence in the lunar anatomy of the bony skull, that strange box of bone that clamps about the developing brain of man, imperiously insisting 'thus far and no farther' to all his possibilities."[160] The idea of a skull as limiting factor in human brain development ominously suggests that a larger-brained species may supersede humanity as the dominant life form on Earth, a possibility Wells explored in *The War of the Worlds* and *The Science of Life*.[161] While the Selenites display no imperialistic desire to colonize Earth, the very existence of such intelligent non-human beings constitutes an implied threat to late-Victorian anthropocentric hubris.

Like the insects they resemble, the Selenites live in a hive-like community where each creature has its function and knows its place. No Selenite is able to function outside his own particular discipline, as a result of innate inclination plus "education and surgery."[162] A Selenite mathematician, for instance, is trained from birth to disregard all other pursuits:

> His brain grows, or at least the mathematical faculties of his brain grow, and the rest of him only so much as is necessary to sustain this essential part of him. At last, save for rest and food, his one delight lies in the exercise and display of his faculty … his brain grows continually larger, at least so far as the portions engaging in mathematics are concerned; they bulge ever larger and seem to suck the life and vigor from the rest of his frame. His limbs shrivel, his heart and digestive organs diminish, his insect face is hidden under its bulging contours.[163]

While he may be perfectly adapted to his lunar society, this Selenite mathematician would horrify educational reformers like Bain and Brown-Séquard, who argued for well-rounded early training in order to promote hemispheric balance.

As Wells's depiction of this unbalanced mathematician shows, the perils of cerebral overdevelopment are social as well as physical and psychological. Physically, the Selenite mathematician's enormous head has developed at the expense of his other body parts, as per Lamarck's first law. Victorian physiologists located higher intellectual functions in the frontal lobes of the brain, perhaps explaining why the alien's bulging forehead obscures his face.[164] Psychologically speaking, this extraterrestrial resembles a monomaniac or even an idiot savant. Given Wells's great admiration of Jonathan Swift, it is no accident that the Selenite's behavior recalls the Laputans, who can only be roused from mathematical speculation when their servants rap them on the mouth and ears.

Socially, the Selenite mathematician seems narcissistic, not to mention perilously isolated from those around him. He lives only to show off his faculty for mathematics. Moreover, Cavor relates that "[h]e seems deaf to all but properly enunciated problems. The faculty of laughter, save for the sudden discovery of some paradox, is lost to him; his deepest emotion is the evolution of a novel computation."[165] For this Selenite, at least, the overdevelopment of mathematical faculties has led to a curious narrowing and flattening-out of emotional experience, as predicted in "The Man of the Year Million": "irrational fellowship … will give place to an intellectual co-operation, and emotion fall within the scheme of reason."[166]

Other Selenites of the "expert class" to which the mathematician belongs are likewise antisocial, even hostile toward those who do not share their monomaniacal interests. They are also surprisingly self-centered for members of a social collective, being "rapt in an impervious and apoplectic complacency from which only a denial of their erudition can rouse them."[167] Phi-oo, the linguistic expert who serves as Cavor's translator, describes the curious habits of a Selenite artist: "M'm – m'm – he – if I may say – draw. Eat little – drink little – draw. Love draw. No other thing. Hate all who not draw like him. Angry. Hate all who draw like him better. Hate most people. Hate all who not think all world for to draw. Angry. M'm."[168] The alien artist's irascibility and fragile ego seem particularly reminiscent of Griffin, another misanthropic expert who refuses to cooperate with other members of his intellectual community.

The physical and social problems experienced by these lunar geniuses are interrelated, since both stem from their remarkable cerebral overdevelopment. The unwieldy shape and size of the lunar intellectuals makes necessary the existence of a servant class of ushers and bearers whose sole function is to "replace the abortive physical powers of these hypertrophied minds." Cavor relates, "Some of the profounder scholars are altogether too great for locomotion, and are carried from place to place in a kind of sedan tub, wabbling [*sic*] jellies of knowledge that enlist my respectful astonishment."[169] Here again, Wells's inspiration likely came from the insect kingdom. Charlotte Sleigh relates how "[a]nts … were well known for their adoption of slaves, and a number of worker-less races had arisen which were incapable of feeding themselves." Swiss researchers conducted experiments upon such ants, proving their inability to eat without assistance: "Even if honey and meat were placed before *Polygerus* workers, they wasted away without their slaves, 'beg[ging] their companions in vain for regurgitation.'"[170] This pitiful example underscores the vulnerability of Selenite experts who depend on their retinue of servants for transportation. These aristocrats' cerebral overdevelopment also has broader implications for the social fabric, which is hierarchical as well as highly specialized. Indeed, the utter dependency of the intellectual Selenites upon the lower classes ominously resembles the grotesquely symbiotic relationship of the Morlocks and Eloi in *The Time Machine*, a novel in which class differentiation has unintentionally disastrous results.

The novel's rather ludicrous climax consists in Cavor's encounter with the Grand Lunar, the Selenite ruler whose massive brain provides the consummate instance of cerebral hypertrophy in Wells's fiction. After ascending several staircases and a series of grand hallways, Cavor is finally allowed to view the majestic presence:

> At first as I peered into the radiating blaze, this quintessential brain looked very much like a thin, featureless bladder with dim, undulating ghosts of convolutions writhing visibly within. Then beneath its enormity and just above the edge of the throne one saw with a start minute elfin eyes peering out of the blaze. No face, but eyes, as if they peered through holes. At first I could see no more than these two staring little eyes, and then below I distinguished the little dwarfed body and its insect-jointed limbs, shriveled and white. The eyes stared down at me with a strange intensity, and the lower part of the swollen globe was wrinkled. Ineffectual-looking little hand-tentacles steadied this shape on the throne … It was great, it was pitiful. One forgot the hall and the crowd.[171]

While Cavor is duly impressed by this rational ruler, the reader is left with a haunting impression of his consummate vulnerability. Because the Grand Lunar lacks a skull, his brain is enormous, visible, and unprotected, while his shriveled body is of no use whatsoever. Predictably, this helpless monarch requires an even larger retinue of servants than the aristocratic experts discussed previously. These include "body servants" that sustain and support his oversized brain case, and "shadowy attendants" who periodically mist his gigantic brow with cooling spray.[172] In terms of appearance, this monstrous being lacks even the semblance of a face that might help to humanize him – but one can hardly help feeling sorry for him nonetheless. In a strikingly obvious way, this "great" yet "pitiful" being demonstrates the impractical, even monstrous result of progressive brain evolution according to the Lamarckian model.

By the novel's conclusion, however, the Grand Lunar turns out to be a relatively unsympathetic character. The closing chapter implies that the ruler orders Cavor's death, having decided that human beings are too acquisitive and warlike to be trusted. Selenite society also turns out to be less benevolent than Cavor believed, since lunar peace and prosperity come at the expense of democratic freedoms that British subjects take for granted. For these reasons and others mentioned above, the Selenites have a fair claim to be the least likable of Wells's geniuses, their cooperation and efficiency notwithstanding. But here as elsewhere in his early fiction, Wells occasionally allows us to identify with these monstrous intellectuals, by witnessing their vulnerabilities, their monomanias, and their evident social incapacity and isolation.

A superficial glance at the above novels might suggest that Wells had an extremely negative view of genius, associating it with amorality, deformity, and alienation (whether literal or figurative). These views were part of the late-Victorian literary and scientific climate in which Wells lived, though no one else dramatized the connection between insanity and genius in such a terrifyingly effective manner. But alongside Wells's dystopian views about human intellectual evolution, the author maintained compassion for the frailties of genius and a deep appreciation for the important role of scientific intellectuals in society. Wells also maintained a correspondingly optimistic interest in the utopian possibilities of brain development. Wells declared in 1904 that "the Instrument of Thought … may have undefined possibilities of evolution towards increased range, and increased power."[173]

Later in life, Wells envisioned even broader possibilities for the mental future of humanity. In 1938, he penned a utopian scheme entitled *World Brain* in which he proposed that "the scientists, technicians and artists, the specialists in all fields, are to be employed in the compilation of a vast, and continually updated world encyclopaedia which will embody the collective wisdom of the world's best brains on every conceivable issue."[174] Wells hoped that this world-encyclopedia or world-brain would serve as a resource for governments all over the globe. Moreover, Wells subtitled his 1934 *Experiment in Autobiography* "Discoveries and Conclusions of a Very Ordinary Brain, since 1866." In this eccentric volume, Wells announced that, "the grey matter of that organized mass of phosphorized fat and connective tissue … is, so to speak, the hero of this piece."[175] The revolutionary move of making a brain, rather than a person, the hero of his personal journey reflects Wells's unrelenting scientific materialism and his firm belief that autobiography should be, perhaps more literally than we are used to thinking, "the story of the contacts of a mind and a world."[176]

## HAVE WE EVOLVED?

What is perhaps most striking about late-nineteenth-century theories about insanity and genius is their persistence in modern culture. The study of correlations between high intelligence and insanity is still a burgeoning field, one which attracts scholars across humanistic and scientific disciplines. Humanists continue to probe the literature of earlier decades in order to explore connections between creativity and madness – in fact, this chapter could serve as one example of such an investigation.[177] Meanwhile, recent scientific research has suggested that there is a strong correlation between artistic creativity, especially creative writing talent, and mood disorders such as bipolar illness or depression.[178] Anecdotal evidence suggests that there may be an equally strong correlation between mathematical creativity and schizophrenia, although no formal research on this topic has been completed.[179] These recent studies are more scientifically rigorous than their nineteenth-century predecessors, using a greater variety of measurable data and more sophisticated data-collection methods. Yet these studies test nearly the same premises laid out by Lombroso, Galton, Nisbet, and other Victorian authors. The idea that genius is somehow associated with insanity is alive and well, and is now supported by more scientific evidence than ever before.

Are we still afraid of geniuses, just like our late-Victorian predecessors? The numerous aliens and mad scientists in recent fiction, films, and television programs would seem to suggest this possibility. Mad geniuses still figure prominently within science fiction, a genre descended from Gothic romances of the eighteenth and nineteenth centuries.[180] But the many humorous incarnations of this figure over the past few decades – such as the title character in *Young Frankenstein* (1974), Dr. Evil in *Austin Powers* (1997), or Dr. Frank N. Furter of *The Rocky Horror Picture Show* (1975) – suggest a growing familiarity and comfort with the concept of the mad scientist. Meanwhile, the ubiquity of large-headed, big-eyed aliens with spindly bodies is a curious legacy of Wells's work, evident not just in Steven Spielberg's blockbuster adaptation of *The War of the Worlds* (2005), but in a broad range of pop-culture phenomena. Nearly all large- and small-screen extraterrestrials boast some variation on this stereotypical alien anatomy – think of E.T. (from the 1982 film of the same name), the alien invaders from Tim Burton's *Mars Attacks* (1996), or even Marvin the

Martian from Looney Tunes. But while some of these large-headed aliens are malevolent, others are approachable, comical, even cuddly.

Because these aliens and mad scientists have become such familiar stock figures in popular culture, it is easy to overlook their historical significance as the most highly visible survivals of late-Victorian mad genius theories. The very ubiquity of such characters, whether humorous or terrifying – combined with the resurgence of scientific studies confirming the link between intellectual talents and certain mental illnesses – suggests that the popular association between madness, genius, and alienation has survived into the present day, even if it has softened somewhat since the Victorian period. Current sociological research supports this idea. In one recent educational project conducted by the United States Department of Energy, researchers asked American seventh-graders to describe and draw their idea of a scientist. Predictably, some described scientists as "crazy," "weird," or "liv[ing] in their own world." Several students drew pictures of large-headed bald men who vaguely resembled the sketch of Wells's future humans in *Punch*.[181] These results are consistent with a range of sociological studies of antiscientific stereotypes conducted over the last half century.[182] Though amusing to recount, findings like these suggest a most serious point: that our attitudes toward genius, particularly scientific genius, have not progressed significantly beyond those of our Victorian predecessors.

CHAPTER 5

# *Marie Corelli and the neuron*

In stark contrast to Robert Louis Stevenson, Bram Stoker, and H. G. Wells, who used romances and Gothic novels to dramatize disturbing, atheistic consequences of cerebral localization, Marie Corelli's bestselling romances creatively envisioned the miraculous possibilities opened up by improved understanding of human brain function. For Corelli, the transcendent potential of neuroscience hinged on the capacity of human nerve cells to receive, store, and conduct electricity, a topic that was the focus of lively scientific inquiry during the late nineteenth and early twentieth centuries. In her 1918 novel *Young Diana*, for instance, the eponymous heroine gains youth, health, and immortality by suffusing her neurons with electricity. After eccentric scientist Féodor Dimitrius immerses Diana in a bath of radioactive fluid for four days, she transforms from an aging spinster into a ravishing, immortal goddess. In the wake of her metamorphosis, Diana gloats: "A goddess – a goddess! … Young with a youth that shall not change – alive with a life that shall not die! Out of the fire and the air I have absorbed the essence of all beauty and power!"[1] While these plot elements fall within the realm of science fiction, *The Young Diana*, like Corelli's earlier compositions, remains firmly rooted in the romance tradition. This example thus demonstrates how Corelli reinvigorated the romance form by liberally intermixing elements of *fin-de-siècle* neuroscience. In so doing, she aimed to reconcile scientific materialism with the imaginative spirituality she viewed as the core of the romance, thereby providing world-weary readers with spiritual solace and renewed vigor. The above example also suggests how Corelli promoted and popularized aspects of cerebral localization theory – such as the controversial neuron doctrine – in the service of the heterodox religious philosophy laid out in her novels. But in order for Corelli's unique fusion of science and spirituality to succeed, she had to willfully *misunderstand* certain tenets of localization, including how neurons actually work. It is unclear whether Corelli's misinterpretation of neuron doctrine stemmed from her lack of scientific education, from the general lack of scientific consensus about neuronal function, or from a selective reading of scientific and pseudoscientific articles on the functions of the brain. In any case, Corelli's fiction, like that of Grant Allen, demonstrates that even apparent champions of cerebral localization could not fully embrace the latest neurological developments – at least not without some adaptation or revision of neurological ideas themselves. This chapter will explore the curious means by which one woman's revision of neuron doctrine helped her readers make a tenuous, temporary peace with the latest neurological developments.

Corelli's ready adoption of late-Victorian neurological concepts in her fiction might seem surprising, given her lack of formal scientific education.[2] But Corelli's status as autodidact probably encouraged her to adapt and manipulate scientific ideas in ways that actual scientists might not – and even extrapolate from current scientific ideas to forecast future technological possibilities. Corelli's talent for scientific "predictions," in fact, rivaled that of more renowned science fiction writers like H. G. Wells or Jules Verne. In various novels, Corelli anticipated such developments as wireless telegraphy, X-rays, biological weapons, and the atom bomb.[3] These predictions reached an extremely wide audience, as her biographer Brian Masters attests:

> While Queen Victoria was alive, Miss Corelli was the second most famous Englishwoman in the world; afterwards, there was no one to approach her. At least half of her books were world best-sellers. About 100,000 copies were sold every year. Her nearest rival was Hall Caine, who sold approximately 45,000 a year … In comparison, H. G. Wells could boast only 15,000 a year.[4]

By some accounts, Corelli's sales actually peaked in the years following Victoria's death. Biographer Annette Federico writes, for instance, that "in 1906 *The Treasure of Heaven: A Romance of Riches* achieved a first-day record of 100,000 copies."[5] The author's massive popularity waned considerably after the beginning of the First World War, when she was falsely accused of food hoarding, but even her last few novels likely had a substantial readership of loyal fans.[6] Many of Corelli's readers were middle-class women like herself, but she also counted luminaries like Oscar Wilde, Prime Minister William Gladstone, Queen Victoria, and the Prince of Wales among her admirers.[7] Corelli was quite right to suggest that her romances "touched a chord somewhere in the great nerves of humanity," as she put it, using neurological metaphors to explain her own success.[8]

Corelli's embrace of cerebral localization theories, particularly the neuron doctrine, played a role in her meteoric rise, perhaps because she adopted these neurological concepts in order to buttress a worldview as different as possible from the materialist perspective of Victorian science. Her early psychical novels, particularly those constituting the so-called "Heliobas Trilogy" – *A Romance of Two Worlds* (1886), *Ardath* (1889), and *The Soul of Lilith* (1892) – lay out a theological belief system whose central tenet is electrical communication between nerves. The "Electric Creed" espoused in these fictions and later works like *The Life Everlasting* (1911) equates electricity with heat, light, and love and depicts neural networks as channels of telepathic communication with Christ.[9] This Creed proved so compelling to readers that it became the foundational doctrine of a small church in the United States.[10]

While recent critics have discussed Corelli's interest in Victorian thermodynamics, psychology, telegraphy, and Edison's electrical discoveries, so far no one has probed her engagement with neurology, despite the innumerable references to brains, brain waves, and brain cells in her fiction.[11] In fact, Corelli's Electric Creed incorporates late-Victorian theories of electric nervous communication, particularly Gustav Fritsch and Eduard Hitzig's 1870 discovery that the cerebral cortex could be stimulated electrically. Their findings suggested that electrical signals might enable communication between different parts of the brain and allow the brain to communicate with the body. Corelli took these discoveries in some frankly occult directions. She speculated that electrical communication might take place not merely within individual brains, but also between *different* brains, making telepathic communication between men and even spiritual entities scientifically plausible (perhaps even empirically demonstrable). For Corelli, therefore, neural pathways became direct routes to other human consciousnesses and to the Godhead itself.[12]

While Corelli co-opted neurological theories that suited her beliefs, such as the theory of electrical nervous communication, she also gradually embraced others (like the neuron doctrine) that heretically suggested a multiplicity of cells and selves. In *Wormwood* (1890), *The Soul of Lilith*, and many subsequent novels, Corelli frequently refers to brain cells, thereby touching upon one of the greatest controversies stemming from cerebral localization research. The existence of brain cells was widely accepted by scientists as early as the 1850s, although mid-century researchers had an imperfect understanding of their structure and function. They debated, for instance, whether "nerve fibers" (what we now call axons and dendrites) touched, forming a web or reticulum, or whether nerves communicated with each other in some other fashion.[13]

The dawning of Corelli's interest in brain cells coincided with the importation of Santiago Ramón y Cajal's neuron doctrine to Britain in the early 1890s.[14] Cajal argued for the "neuron" as the anatomical and physiological unit of the nervous system, and contended that impulses pass from one neuron to another unidirectionally along axons and dendrites.[15] Though most mainstream scientists accepted the neuron doctrine, they debated the details of the theory and its implications. Scientists wondered, for instance, how nerve impulses pass between neurons. Since late-Victorian microscopes were not powerful enough to view the synaptic gap, it was impossible to observe directly how neurons communicated.[16] A few scientists rejected the neuron doctrine altogether, preferring the mid-Victorian concept of the brain as reticulum or network of interconnected nerve fibers. For instance, Italian neurologist Camillo Golgi denounced neuron doctrine and cerebral localization more generally when he received his Nobel Prize in 1906. He shocked the scientific community by cleaving to the older, holistic theory that the brain was a "diffuse nerve network."[17]

While Corelli likely did not know the intricacies of the debates surrounding neuron doctrine, she introduced the concept of the brain cell into her fiction as early as 1890. In her novel *Wormwood*, the absinthe-addicted protagonist is told, "your brain cells are still heavily charged with … poison."[18] Though the phrase "brain cell" had been used in scientific literature since the early nineteenth century, it was quite rare at this time for a literary author to use this terminology.[19]

Corelli revisited the concept of the brain cell in most of her subsequent works, demonstrating her continued interest in neurology and its potential application to spiritual life. In *The Soul of Lilith*, Corelli describes "curious little brain-particles that lie in their various cells."[20] By the time she wrote her blockbuster hit *The Sorrows of Satan* in 1895, Corelli had an improved but still imperfect understanding of the brain cell: "The brain cells are atoms, and within these, are other atoms called memories, curiously vital and marvelously prolific!"[21] Intriguingly, Corelli's *misunderstanding* of the neuron proved to be her most ideologically significant contribution to the debates surrounding cerebral localization. By envisioning the neuron as a storehouse for electricity and memory as well as a fundamentally distinct unit of the nervous system, Corelli was able to incorporate the concept of the brain cell into her popular Electric Creed, thereby making the controversial neuron doctrine palatable to the masses.

Corelli's most significant achievement as a novelist, however, lies in her unique manner of fusing the romance with neurology. Whereas Stevenson and Wells wrote Gothic romances that jarred readers' nerves, Corelli aimed to soothe and uplift her audiences by helping them reconcile religious faith and scientific progress. Moreover, Corelli arguably took the romance more seriously than Stevenson, Andrew Lang, and others who construed the primary function of the genre as escapism and wish-fulfillment. In Stevenson's words, the goal of romance is "the realization and the apotheosis of the day-dreams of common men." A romance should "satisfy the nameless longings of the reader" and "obey the ideal laws of the day-dream."[22]

Though Corelli's novels contain ample doses of wish-fulfillment, she conceived of her romances as more than just escapist fantasies – they also served a therapeutic function. She offered up her romances "out of love and pity for suffering humankind," hoping to transmit "peace and comfort" of mind to despairing readers.[23] She modeled the healing process she envisioned for her readers in her protagonists' life stories. Corelli's readers vicariously recaptured youth and vitality through romances whose heroines underwent similar transformations. Like Diana, whose shriveled nerve cells swell with renewed electricity during her radioactive bath, Corelli's readers restored their depleted nervous energies by temporarily immersing themselves in a fictional realm of beauty and spiritual uplift. Corelli carefully guarded readers from jarringly realistic notes, arguing instead that imaginative fiction should present "the *idealisation* of human thought into *ideal* language," in order to "break open the close walls of our earthly prison house and let a glimpse of God's light through."[24]

Judging from her fan mail, Corelli's romances often succeeded in strengthening readers' faiths and improving their mental and physical well-being. One Anglican clergyman wrote to Corelli that her vastly popular first novel, *A Romance of Two Worlds*, rescued him from suicide:

> Once I believed in the goodness of God – but of late years the cry of modern scientific atheism, "There is *no* God," has run in my ears till my brain has reeled at the nothingness of the Universe ... I began to read, and never left it till I had finished the last page – then I knew I was saved.[25]

Another writer insisted that after reading *Romance*, "I feel now as if I had, like a leper of old, touched the robe of Christ and been healed of a long-standing infirmity."[26] In her "Appendix" to *Romance*, Corelli avowed that "scarcely a day passes without my receiving more of these earnest and often pathetic appeals for a little help, a little comfort, a little guidance."[27]

In positing a medicinal function for romantic narratives, and even successfully "curing" readers of spiritual and psychological ills, Corelli exceeded even the strongest arguments in favor of romance put forth by Stevenson, Lang, or Wells.

## NERVOUS PATHWAYS TO GOD

In contrast to Stevenson's romantic adventure tales, which primarily appealed to young males, Corelli's neurological romances possessed special attractions for female readers. At a time when women were widely considered more "nervous" than men and hence more prone to mental illness, Corelli's fictional heroines capitalize on their nervous energies to achieve personal autonomy and spiritual insight. Some of her female protagonists even travel through space and time by harnessing the electrical forces latent in their nerve cells. Though Corelli was not a feminist in the traditional sense – she ridiculed the "New Woman" of the 1890s and initially opposed the Women's Suffrage movement – her fiction steadfastly promoted women's intellectual and artistic freedom.[28] In a radical reversal of Victorian gender norms, Corelli posited that women's perceived weakness (excessive nervousness) could become a means of achieving personal freedom. More radically still, Corelli suggested that women's spiritual energies, when properly channeled, could facilitate direct communication with God.

For instance, the heroine of *A Romance of Two Worlds* cures her own depression and views heaven itself, all by harnessing her internal stores of nervous electricity. This work is where Corelli first introduces her popular Electric Creed. Since this novel is now largely unknown, a brief summary of its unconventional plot seems in order here. The unnamed protagonist is a female concert pianist suffering from insomnia and depression due to overwork. In these respects, she resembles Corelli herself, who pursued an ultimately unsuccessful career as a professional pianist in the mid 1880s. A nineteenth-century physician would probably have diagnosed the protagonist with nerve exhaustion or neurasthenia, and prescribed a cure consisting of rest and overfeeding.[29]

Corelli's protagonist rejects this confining regimen. Instead, she consults a homeopathic spiritualist guru named Heliobas, who adheres to a religion known as the "Electric Creed of Christianity."[30] The heroine recovers her health and strength through a regime of revitalizing herbal tonics, electrotherapy, and prolonged soul-searching. Meanwhile, she becomes more acquainted with the Creed of Heliobas, who uses his electric powers to free the protagonist's soul from her body. With her soul thus liberated, she traverses the galaxy, viewing life on other planets. She even sees heaven itself, which looks like "[a] Sphere … marvellous and dazzling; a great globe of opal-tinted light, revolving as it were upon its own axis, and ever surrounded by that scintillating, jewel-like wreath of electricity, whose only motion was to shine and burn within itself for ever."[31] The heroine is permitted to view these exalted regions because she is spiritually prepared; she has already "cultivat[ed] and educat[ed] the electric force within" (*RW* 242).

Corelli's unorthodox heaven is, like the heroine's body, based on an internal economy of energy. Like a star, Corelli's heaven is a constant, brilliant source of light and heat. It is filled with divine spirits who radiate God's love to people on Earth. Moreover, heaven is incredibly vast, "so large that multiplying the size of the sun by a hundred thousand millions, no adequate idea could be formed of its vast proportions" (*RW* 233). When the heroine's soul eventually leaves these astral climes and returns to her body, she emerges as a strong defender of the Electric Creed. In the course of her exhilarating journeys, she also learns to project her nervous energies in order to improve her environment, health, and spiritual well-being.

Corelli further describes her Electric Creed in the "Author's Prologue" to the 1887 edition of *Romance*:

> The idea of Eternity [is] depicted in the "Electric Ring encompassing God's World" … The Ring is … *perpetually creative and perpetually absorbent.* Planets are from time to time drawn within it and cast forth from it, and of this tremendous electric Force there can be no end, inasmuch as it is the outer circle or atmosphere of the Central Planet of all planets wherein the Creator has His being.[32]

In Corelli's unusual cosmology, "[a]ll the wonders of Nature are the result of *light and heat alone*," and light, heat, and electricity are equated with God's love (*RW* 390). Eschewing the traditional idea of Hell, Corelli suggests that erring souls must be reincarnated repeatedly so that they can learn from and overcome past mistakes (here she draws on elements of theosophy and late-Victorian popular understandings of Buddhism). But since memories carry over into future lives, each errant soul remains tormented by the recollection of misdeeds.

Corelli modified the Creed somewhat in later romances such as *The Life Everlasting*, substituting radium for electricity. Still, the basic tenets of her philosophy remained the same. In this later novel, another nameless heroine resembling Corelli goes yachting off the coast of Scotland with a millionaire and his neurasthenic daughter. They encounter an Egyptian sage named Rafel Santoris, who is a disciple of Heliobas's successor, Aselzion. Santoris practices a kind of "Spiritual science" that gives him the secrets to eternal youth, health, and immortality (*LE* 133). Discovering that Santoris is her soul mate, the protagonist embarks on a course of study with Aselzion in order to master these secrets for herself. Like the heroine of *Romance*, she learns to channel her internal energy reserves to achieve eternal youth and communicate with angels. When she successfully completes her studies, the lovers reunite and live in perfect health, wealth, and happiness aboard Santoris' electric-powered yacht.

Intriguingly, Corelli felt the need to emphasize her personal connection to these fantastic narratives. In her "Author's Prologue" to the *Romance*, she declared, "I can only relate what I myself have experienced," whereas in *The Life Everlasting* she more carefully distanced herself from the events of the story: "I am not the heroine of the tale – though I have narrated it (more or less as told to me) in the first person singular" (*LE* 34).[33] To her close friends, Corelli admitted that the prologue to *The Life Everlasting* accurately represented her own religious beliefs. In a 1923 letter to journalist John Cuming Walters, written a week before her death, she wrote: "I am sending you *The Life Everlasting*. If you will read the 'prologue' only, you will grasp my 'faith.'" She had been developing and following this faith "all my life," she told Walters.[34]

Corelli's Electric Creed was unique in its emphasis on nerves and nervous energies. In other respects, however, Corelli's unusual faith owed much to heterodox spiritual traditions such as Rosicrucianism, Christian Science, and Victorian spiritualism. The story of Corelli's neurological romances would be incomplete without acknowledging her debt to these traditions. Her novels repeatedly depict mental telepathy between human beings, angels, and God, not to mention materialization of spirit forms. Bizarre as these phenomena may seem to the modern reader, late-Victorian audiences likely found themselves on familiar ground. Corelli began writing toward the end of the "golden age of English spiritualism," which lasted roughly from the 1860s through the 1880s.[35] Romances based on spiritualist themes were relatively common during these years, particularly in America, where Elizabeth Stuart Phelps's *The Gates Ajar* (1868) became a runaway bestseller.

This vogue for spiritualism was motivated by Victorians' loss of religious faith in the face of scientific discoveries that increasingly called into question man's place in the universe.[36] Spiritualists – who were often practicing Christians, if not strictly orthodox ones – sought to accommodate their beliefs to the prevailing materialist intellectual climate by investigating spirit phenomena they could see, hear, and measure with varying degrees of objectivity. Spiritualists ultimately wished to prove the existence of the immortal soul, which seemed to be verified by the appearance of apparitional visitors from beyond the grave (a common occurrence at séances). Though precise numbers of Victorian spiritualists are difficult to ascertain, Janet Oppenheim estimates that somewhere between ten thousand and one hundred thousand Britons embraced spiritualist practices.[37] Far from constituting a "lunatic fringe of their society," spiritualists were situated "squarely amidst the cultural, intellectual, and emotional moods of the era."[38]

Despite Corelli's biting denunciations of spiritualism, theosophy, and séances, which she condemned as "contemptibly trivial in character, and vulgar, when not absolutely ridiculous," both her contemporaries and recent critics have observed that the Electric Creed contains strong overtones of spiritualism and frequent references to the theosophical belief in reincarnation.[39] Corelli complained in 1894 that "critics are 'down upon me' because I write about the supernatural."[40] Realizing that she was in danger of being stereotyped as a spiritualist writer, Corelli deliberately varied her novelistic output to include more traditional thrillers and romances: "I decided to … change my line of work to lighter themes, lest I should be set down as 'spiritualist' or 'theosophist,' both of which terms have been brought into contempt by tricksters" (*LE* 28–9).[41]

Corelli likely refrained from identifying herself with spiritualism, mesmerism, and related phenomena for several reasons. Most obviously, she hesitated to affiliate herself with occult trends that garnered ridicule in the popular press. She aspired to become a respectable author rather than merely a popular one, a goal that largely eluded her due to scathing critical reviews of her books.[42] Moreover, Corelli's stepfather, journalist Charles MacKay, derided mesmerism in his *Memoirs of Extraordinary Popular Delusions and the Madness of Crowds* (1841), suggesting that her early upbringing was hostile to the kinds of occult phenomena that would later interest her.[43] Nonetheless, Corelli was well aware that spiritualists were her most eager readers. Shortly before the appearance of *Ardath*, Corelli confessed that "spiritualists … are all waiting for my new book, as cats for a poor little mouse."[44] After Corelli's death, at least five books were written by authors who claimed to be guided by her spirit. In America, where spiritualism enjoyed the greatest popularity, Corelli even had a town named after her (Corelli City in Colorado).[45]

Like many spiritualists, Corelli felt a strong aversion toward Victorian scientific atheism, combined with resentment of perceived scientific arrogance. She dedicated her wildly popular novel, *The Mighty Atom*, "To *those self-styled 'progressivists,'* who by precept and example assist the infamous cause of *education without religion* … denying to the children in board-schools and elsewhere, *the knowledge and love of God* as the true foundation of noble living."[46] Such progressivists, Corelli wrote, "*are guilty* of a worse crime than murder."[47] Here and elsewhere, Corelli depicted scientists and intellectuals generally as arrogant and unfeeling men whose materialistic worldview was misguided at best, malevolent at worst. Many of her readers presumably shared these feelings, as they clung to their spiritual convictions in an increasingly skeptical intellectual climate.

Like Bram Stoker, then, Corelli's intellectually daring theories stemmed partly from reactionary sympathies. But her opposition to scientific materialism should not obscure her abiding interest in scientific ideas themselves, especially insofar as they could be used to buttress religious faith. In the Heliobas trilogy and elsewhere, Corelli invoked new scientific research about nerve force and neurons in order to reconcile Victorian science with a certain type of heterodox, spiritualist Christianity. She explains in her "Author's Prologue" to the *Romance*: "the light of Science must be brought to bear on the New Testament, in which its glorious pages will grow bright with hitherto unguessed mystical meanings."[48]

By the early twentieth century, Corelli had also incorporated tenets of Rosicrucianism and Christian Science into her Electric Creed. Like Corelli, followers of these faiths attempted to reconcile scientific principles with spiritual teachings. Moreover, both Rosicrucianism and Christian Science emphasized the role of personal autonomy in achieving health and spiritual well-being, a welcome message for women seeking greater control over their destiny. Rosicrucianism, which traces its roots to ancient Egyptian mysticism and medieval alchemy, teaches that each person must undertake a spiritual quest in order to learn the secrets to life, health, and psychic powers. This spiritual formula resembles the plotlines of several Corelli novels, especially *Romance*, *The Life Everlasting*, and *Ardath*. Rosicrucianism explicitly surfaces in Corelli's *Young Diana*, where eccentric scientist Dimitrius explains that "[t]he Rosicrucians have come nearer than any other religious sect in the world to the comprehension of things divine."[49] The official website of the Rosicrucian Order even lists Corelli among a number of famous people associated with the sect.[50]

Similarly, echoes of Christian Science can be found in Corelli's work, especially her belief that the mind can control the health of the body, and that life's tribulations are "born of our own wrong thinking, and are not sent from God" (*LE* 98). Christian Science, founded by New England homeopath Mary Baker Eddy in the early 1870s, emphasized "the triumph of Spirit, Mind, over matter," especially the power of the mind to overcome bodily illness.[51] This new religion peaked in popularity in the early twentieth century, due partly to public dissatisfaction with mainstream health care and women's desire to control their spiritual and medical destinies.

From its beginnings, Christian Science was a largely female movement rooted in spiritualist, theosophical, and Swedenborgian traditions.[52] Like Eddy, Corelli emphasized the gulf separating "the perishable materialism of our ordinary conceptions of life, and the undying spiritual quality of life as it truly is" (*LE* 26). The protagonist of *The Life Everlasting* repeatedly denies practicing Christian Science, thereby acknowledging how easily her views might be confused with those of Eddy and her followers (*LE* 100, 110). Indeed, Corelli resembles Mary Baker Eddy and the more powerful female practitioners of New Thought (a related movement emphasizing mental healing) in that she was an influential female founder of a religious doctrine who believed that spirit trumped matter.

The modern reader may well wonder whether Corelli could be best described as an idiosyncratic religious visionary along the lines of William Blake, a charismatic religious leader similar to Eddy, or a skilled cultural broker who deftly interpreted and repackaged trendy social, spiritual, and scientific references in a manner calculated to supply maximum popular satisfaction. Though Corelli's dedication to her unusual creed was undoubtedly sincere, I suspect her gifts were primarily of the latter sort. This conclusion seems particularly likely when one considers her obvious business savvy and the wide appeal of her unusual productions, which drew on numerous popular faith movements of especial interest to Victorian women.

Corelli's merging of science and spirituality grew out of her shrewd business instincts as well as personal conviction. She wrote to her first publisher, George Bentley, that "people appear to revel in and gloat over anything that has to do with an admixture of science and religion."[53] Corelli knew that her books filled a profound and widespread need for reassurance that science and "eternal things" were not incompatible. But she was not just shrewdly pandering to her audiences, since she shared her readers' psychological longings. "One sighs for the old grand days of trust in God," Corelli told Bentley. Echoing the language of her fan mail, she wrote, "It is enough to make the strongest brain reel, to force it to speculate on the blank horror of an utterly *purposeless* Universe."[54]

As Corelli well knew, a certain group of scientists likewise sought to reconcile science and "eternal things." Her novels owe an intellectual debt not only to the abovementioned religious traditions, but also to the work of the Society for Psychical Research (SPR). A group of Cambridge professors founded the SPR in 1882, four years before Corelli's first novel was published. The Society's express purpose was to discover whether there was any scientific basis to occult phenomena such as haunted houses, ghosts, Ouija boards, and mental telepathy. Perhaps surprisingly, these outré subjects attracted the interest of many illustrious researchers. At the height of the SPR's fame, its members included Pierre Janet, Cesare Lombroso, G. Stanley Hall, Sigmund Freud, and William James.[55] The involvement of such luminaries suggests that many Victorians viewed the investigation of occult phenomena as a legitimate scientific enterprise.

Though some SPR members were motivated by sheer scientific curiosity, a fair number (including future SPR Presidents Frederic Myers, Oliver Lodge, and William Barrett) were driven, like Corelli, by a desire to prove the existence of an immortal soul. These spiritual yearnings often interfered with strict scientific objectivity. In the 1880s, some SPR members felt that they had demonstrated the existence of certain occult phenomena beyond a reasonable doubt. Edmund Gurney, Frederic Myers, and Frank Podmore confidently declared in *Phantasms of the Living* (1886) that "telepathy – the supersensory transference of thoughts and feelings from one mind to another – is a fact in nature," supporting this claim with considerable experimental and anecdotal evidence.[56]

Around the same time, physicists Oliver Lodge and William Barrett conducted their own experiments in thought-transference. Lodge published an article in *Proceedings of the Society for Psychical Research* in 1884 in which he compared electric induction to mental telepathy. (Electric conduction is the flow of electricity along wires, whereas induction is "the phenomenon that produces an electrical charge in an adjacent or proximate circuit, apparently acting across empty space.")[57] Lodge wrote:

> That the brain is the organ of consciousness is patent, but that consciousness is located in the brain is what no psychologist ought to assert: for just as the energy of an electric charge, though apparently on the conductor, is not on the conductor, but in all the space around it … so it may be that the sensory consciousness of a person, though apparently located in the brain, may also be conceived

of as existing like a faint echo in space, or in other brains, though these are ordinarily too busy and pre-occupied to notice it.[58]

Later, Lodge began to think more deeply about the physics of this argument, drawing upon Heinrich Hertz's 1889 experiments with "spark gaps" in circuits. Hertz had shown how a discharge of electricity in one circuit could produce a spark or "sympathetic vibration" in a second circuit located at some distance from the first. In a 1900 lecture, Lodge demonstrated this principle with tuning forks, concluding that "sympathetic vibrations" between brains might facilitate thought-transference.[59] However, just as two dissimilar tuning forks might not respond to one another, two complete strangers are less likely to experience sympathetic vibrations than family members or other persons who feel a "sympathy" between them.[60]

Barrett, who conducted his own experiments with tuning fork resonance in the late 1860s, explained thought-transference as "synchronous vibration" between brains:

> We may … conceive of nervous energy acting by induction across space as well as by conduction along the nerve fibres. In fact, the numerous analogies between electricity and nervous stimuli would lead to some such inference as the above. Or the brain might be regarded as the seat of radiant energy like a glowing or sounding body. In this case, the reception of this energy would depend upon a possibility of synchronous vibration in the absorbing body … so that a distant mental disturbance might suddenly and profoundly agitate particular minds, whilst others might remain quiescent.[61]

Elsewhere, Barrett referred to telepathic induction as "exoneural action of mind," neurological phenomena powerful enough to transcend the confines of the brain.[62]

Psychical researchers like Lodge and Barrett made analogies between telegraph and telephone communication and thought-transference, demonstrating the considerable overlap between science and pseudoscience when it came to communicative "technologies."[63] For instance, Lodge compared thought-transference between "two brains" to "a couple of telephones connected properly by wires. They are sympathetic, and if one is tapped the other receives a shock."[64] Significantly, Lodge not only theorized about telepathy but also helped develop wireless telegraphy in the 1890s, an invention that Corelli's first novel predicted a decade in advance of its achievement.

Although Corelli savaged Oliver Lodge and other psychical researchers in her non-fictional writings, her romances unmistakably display the influence of the SPR. In her "Author's Prologue" to *The Life Everlasting*, Corelli wrote:

> You have at present living among you a great professing scientist, Dr. Oliver Lodge, who, wandering among mazy infinities, conceives it even possible to communicate with departed spirits, – while I, who have no such weight of worldly authority and learning behind me, tell you that such a thing is out of all natural law and therefore *can never be*. (*LE* 29–30; emphasis in original)

Elsewhere she remarked that psychical researchers were suffering from bilateral brain hemispheric imbalance, commonly thought to produce insanity ("The twin lobes of the brain are not acting in union").[65] But Corelli's mockery of Lodge suggests, paradoxically, that she had at least some familiarity with his ideas. Her fiction does nothing to allay these suspicions, particularly her use of the term "sympathetic vibration" (*LE* 103, 170). Corelli's grounds for repudiating Lodge and other psychical researchers may well have been personal rather than ideological. Like many spiritualists, she resented the scientific, skeptical bias of the SPR and the "worldly authority and learning" of its illustrious members, a surprising number of whom, like Lodge, were Fellows of the Royal Society.[66] Later in life, Corelli apparently overcame this prejudice, striking up a friendship with physicist and psychical researcher William Crookes, FRS.[67]

In fact, Corelli had much in common with avowedly spiritualist members of the SPR such as Lodge, Barrett, and Crookes, particularly her faith in telepathy and her desire to prove scientifically the existence of an immortal soul. What Oppenheim says of Lodge could apply equally to Corelli: "in characteristic spiritualist fashion, [Lodge] believed that it was his ongoing mission, his special concern, to reconcile science and religion."[68] Like Corelli, Lodge eschewed the Christian idea of Hell and envisioned a universe in which matter and spirit functioned in harmony. Barrett, meanwhile, revealed his spiritualist sympathies when he plaintively inquired, "May there not be some telepathic inter-communion between the Creator and all responsive human hearts, to some being given the inner ear, the open vision, and the inspired utterance?"[69]

Corelli's novels *A Romance of Two Worlds* (published the same year as *Phantasms of the Living*) and *The Life Everlasting* popularized Barrett's and Lodge's scientific explanation for mental telepathy. Like these psychical researchers, Corelli suggested that electrical conduction via nerves and electrical induction between brains facilitated telepathic communication. Also like Barrett and Lodge, Corelli felt that telepathy was not possible on all occasions, only between individuals whose souls were "placed on the same circle of electricity" (*RW* 107). In *Romance*, she describes the role of conduction in transmitting nerve impulses: "That wondrous piece of human machinery, the nervous system; that intricate and delicate network of fine threads – electric wires on which run the messages of thought, impulse, affection, emotion" (*RW* 63). In *The Life Everlasting*, meanwhile, Corelli presents an explanation of telepathy via electrical induction that strikingly recalls Barrett's and Lodge's theories: "Just as the human voice is transmitted through distance on the telephone wires, so is the Soul's voice carried through the radiant fibres connected with the nerves to the brain. The brain receives it, but cannot keep it – for it is transmitted by its own electric power to other brains" (*LE* 374). Here, Corelli draws an analogy between technologically facilitated communication (like the telephone and telegraph) and nervous communication, as members of the SPR so often did. Laura Otis has explained that comparisons between mechanical and organic communication systems were common throughout the nineteenth century; in fact, the inventors of the telegraph and physiologists studying the nervous system looked to each other for inspiration.[70] The writings of Marie Corelli and members of the SPR present striking examples of these biomechanical metaphors.

But whereas most SPR members thought telepathy was only possible between living humans, and spiritualists argued for the possibility of communication between the living and the dead, Corelli made more sweeping claims. She posited that telepathic connection could be established between human mediums and God, a possibility Barrett had tentatively proposed but never pursued. By contrast, Corelli boldly proclaimed that "God's Cable is laid between us and His Heaven in the person of Christ" (*RW* 279). Communication with extraterrestrials and spirit entities might also occur, she thought. As one character in the *Romance* asks, "Why should not a communication be established, like a sort of spiritual Atlantic cable, between man and the beings of other spheres and other solar systems?" (*RW* 200). (The Atlantic Cable, which was completed in 1866, allowed telegraphic messages to be sent between the United States and Europe.) In short, Corelli's neurological romances deftly wove together the aims of spiritualists, psychical researchers, and heterodox Christians, creating a uniquely populist form of spirituality that especially appealed to her female readership.

## NERVOUS MECHANISMS: THE CIRCUIT AND THE BATTERY

Corelli's Electric Creed responded not just to spiritualists and psychical researchers – whose occult inquiries occupied the fringe territories of respectable science – but also to the most mainstream scientific thinkers of the eighteenth and nineteenth centuries. Her omnivorous mind greedily absorbed concepts from neuroscience, thermodynamics, and electrical engineering, fields that serendipitously overlapped in the late-Victorian period.

But whereas scientific thinking typically proceeds in linear fashion, Corelli's romances used the metaphor of the circle or circuit to describe an ongoing process of energy renewal and spiritual regeneration. Corelli's romance plots frequently circle back to characters' childhoods, or even to previous lives, since her belief in reincarnation allowed her to expand narrative arcs over numerous lifetimes. *Ardath*, for instance, traces the progress of the protagonist's soul over the course of many reincarnations and several millennia. The circularity of Corelli's romances made them a perfect vehicle for dispensing her popular Electric Creed, which described Heaven as an enormous electric ring and posited that individuals exist in different "circle[s] of electricity" (*RW* 107).

Corelli's circular narratives feature numerous references to electric circuits and a related device, the electric battery. Both biomechanical structures played key roles in Romantic and Victorian studies of neuroanatomy. Since Corelli's Electric Creed borrows much from Romantic-era vitalists, we must begin with the enormously influential debates between Luigi Galvani and Alessandro Volta over whether nerves conduct electricity, and whether nervous electricity originates in the body. In 1791, Galvani published *De viribus electricitatis in motu musculari commentarius*, in which he describes how he electrically stimulated frogs' legs to produce twitching movements. The creation of an electric circuit was essential to the experiment, as Sidney Ochs explains:

> [Galvani] found that when frogs were suspended on an iron balcony by a brass hook, with their legs hanging down, contractions occurred when they touched the iron balcony. Imitating the effect of the two metals, he fashioned from them what he referred to as a "metal arc." By means of the arc, muscle contractions were produced when they completed a circuit between the muscle and the rest of the animal.[71]

This famous experiment and a subsequent 1794 publication by Galvani and his nephew, Giovanni Aldini, demonstrated to the satisfaction of many contemporaries that nerves conduct electricity. Galvani also theorized that the cerebrum itself generated an electric fluid that flowed down the spinal cord and into nerves innervating muscles, a view that proved highly influential.[72]

Many scientists, particularly vitalists and *Naturphilosphen*, inferred from Galvani's work that electricity was an essential life force that could revivify the dead. Some researchers used electricity to temporarily reanimate corpses of executed criminals, following the lead of Aldini, who was able to excite muscle contractions in hanged murderer Thomas Forster shortly after his 1803 execution. Such experiments became so popular that the Prussian government passed an edict in 1804 forbidding the use of decapitated criminals' heads for galvanic experiments.[73] Grotesque medical "research" of this nature no doubt influenced Mary Shelley's *Frankenstein* (1818), whose eponymous scientist-antihero harnesses electricity to bring to life a reconstituted corpse.[74]

As the latter example suggests, Galvani's findings inspired the literary intelligentsia, who took a typically Romantic view of electricity as a life-giving and death-dealing power with immense spiritual significance. Corelli, an avid reader of Romantic poets since childhood, may well have imbibed her enthusiasm for electricity through their writings.[75] Johann Wolfgang von Goethe, for instance, described electricity as "the soul of the world."[76] Romantics emphasized the mobility and fluidity of this mysterious force which is independent of gravity and borders and whose shapeless flowing stream resists quantification and particularization. Electricity was equated by some with unmediated and endless communication, and was frequently associated with youth, erotic attraction, and femininity.[77]

However impressive and influential Galvani's findings proved, they did not go unchallenged for long. Beginning in 1792, physicist Alessandro Volta argued that Galvani's experiments did not demonstrate the existence of organic nervous electricity. Volta thought that electrical conduction between dissimilar metals had caused the frogs' legs to twitch, rather than electricity generated by the animals' own bodies. To demonstrate the power of metallic electricity, Volta created the first man-made electrical battery in 1800 using alternating plates of unlike metals. The design of this "artificial electric organ" was based on the natural electric organ of the electric fish, which had been well known since antiquity due to its power of producing electric shocks.[78] Curiously, Corelli knew of the electric fish and its connections to the Voltaic "pile" or "cell." In *Romance*, the heroine reads about "electric organs as they are discovered to exist in certain fish … the arrangement of their parts may be compared to a Voltaic pile. They develop electricity and give electrical discharges" (*RW* 197). Corelli was likewise familiar with the supposed healing powers of these fish, whose shocks were thought to relieve muscle paralysis, headaches, and other ailments.[79]

Corelli's knowledge of electric fish, not to mention her view of electricity as vital force, might lead one to believe that her scientific views were outdated. But in fact, many Victorians were just as enamored of the mysterious powers of electricity as their Romantic predecessors. This is partly due to the discoveries of Hermann von Helmholtz, who ushered in a new era in physics in 1847 with his theory of the conservation of energy, also known as the first law of thermodynamics. This principle states that the total amount of energy in an isolated system remains constant, although it may change forms (for instance, friction turns kinetic energy into thermal energy). The first law also suggests that all forces of nature (mechanical, electrical, chemical, etc.) are "forms of a single, universal energy, or *Kraft*, that cannot be either added to or destroyed."[80] Many Victorians found spiritual significance in Helmholtz's "gospel of energy."[81] Biologist Jacob Moleschott, for instance, declared that "the [scientific] materialists profess the unity of energy and matter, of the spirit, of the body, of God and of the world."[82] Corelli's idea that the Electric Ring surrounding heaven is "perpetually creative and perpetually absorbent" likewise draws on the first law to create an image of spiritual harmony.[83]

As the century progressed, technological applications of the laws of thermodynamics solidified the central importance of electricity in the public eye. In the late 1860s, Werner von Siemens invented the dynamo, which made possible the industrial use of high-voltage electric current in factory operations. By 1882, Thomas Edison could supply power to light bulbs in private homes from an electric power center. These new inventions were displayed at the Paris Electricity Exhibition of 1881 alongside "cables, telephones, [and] the first streetcar with overhead power supply."[84] The Electricity Exhibition captured the public imagination in a manner similar to Galvani's experiments, showcasing electricity as a vital force that was both life-giving and destructive.[85]

These inspiring technological innovations prompted a renaissance of Romantic ideas about electricity, particularly among Victorian spiritualists. Corelli, who lauded electricity as "the wonder of our age" and argued that "electricity is life," was not the only literary author to get caught up in the subsequent excitement (*RW* 85, 291). For instance, French novelist Auguste Villiers de L'Isle-Adam's *Tomorrow's Eve* (1886) featured Thomas

Edison as its protagonist and reprised some of the same electrical creation themes explored in *Frankenstein*.

The late-Victorian period not only ushered in a renewed interest in electricity itself, but also increased public awareness that, despite Volta's claims, electricity was in fact central to the functioning of the nervous system. Physiologist Emil Du Bois-Reymond initiated the modern phase of nerve physiology in the 1840s by establishing the existence of electric nerve currents.[86] In 1850, meanwhile, Helmholtz demonstrated the difference between metallic electricity and nervous electricity by showing that they travel at different velocities, thus clearing up a confusion that had begun with the Galvani–Volta debates. Although Galvani's ideas about the electrical nature of nerve function were primitive, particularly since he erroneously viewed electricity as a fluid, his basic premise had ultimately been proven right by mid-nineteenth-century German physicists.[87]

The role of the brain itself in bio-electrical nervous processes was still shrouded in mystery during the mid-Victorian period. Was the brain a sort of electric battery, as Galvani had proposed? Could the cerebral hemispheres respond to electrical stimuli in the same way as nerves elsewhere in the body? The latter controversy raged until the late 1860s, before which scientists had failed to excite the hemispheres using either mechanical or electrical stimuli (results that must be attributed to faulty experimental techniques or scientific instruments).[88] In 1870, however, German physicians Gustav Fritsch and Eduard Hitzig published their landmark work, "On the Electrical Excitability of the Cerebrum." In this paper, they described experiments in which they had electrically stimulated certain parts of dogs' cerebrums, producing muscular contractions on the opposite side of the body. These findings revolutionized late-Victorian neuroscience by ushering in an era of cerebral localization research and by establishing beyond doubt that the brain responds to electrical stimuli.[89]

Some scientists postulated that the brain not only responds to electricity, but also generates and stores it. The brain's energy, it was thought, could be expended in the form of willpower. For instance, chemist Paul Traugott Meissner argued in the 1830s that the will operated by means of electricity that emanated from the brain itself.[90] Later in the century, English physiologist William Carpenter's highly influential *Principles of Mental Physiology* (1874) described the brain as an energy source powering a complex electrical circuit.[91] Corelli was evidently thinking along these lines in her fiction, for instance, when Heliobas encourages the heroine of

*Romance* to "educat[e] your Will to certain height of electric command" (*RW* 243).

The idea of the brain as battery powering the body's nervous circuitry also surfaced in America from the late 1860s onward, where it became the basis for the new and trendy neurasthenia diagnosis. Neurasthenia, which was first described by American neurologist George Beard in 1869, was a broadly defined nervous ailment encompassing such diverse symptoms as depression, rashes, insomnia, and migraines, all of which supposedly resulted from "deficiency or lack of nerve force."[92] This new diagnosis combined recent discoveries about the electrophysiology of the brain and nervous system – particularly Helmholtz's discoveries about the velocity of nervous electricity – with folk beliefs about the limited quantity of available energy in the human body.[93] Corelli's novels contain a plethora of neurasthenic characters, demonstrating her familiarity with the disorder and its treatments.

The development of the neurasthenia diagnosis was the logical extension of Helmholtz's law of conservation of energy to the human nervous system.[94] Doctors treating neurasthenics considered the body a vast reservoir of energy. They argued that this energy could easily become depleted due to excessive expenditure, inadequate supply, or "defective organization in the energy reservoir."[95] For instance, Beard likened the neurasthenic body to "an electric battery that does not supply very much more electric force than is needful for the use to which the battery is put."[96] The primary treatment Beard recommended for neurasthenics was electrotherapy, in which electric charges were applied to the body or head using a galvanic battery. This treatment supposedly restored the body's depleted nervous energy.[97]

Later researchers studying neurasthenia suggested that the body's nerve force was generated in brain cells rather than diffused throughout the body. This view reflected neurologists' growing but still imperfect knowledge of the electrophysiology of the neuron. This subject was not well understood until the mid-twentieth century, when the energy produced by neurons was found to be a byproduct of chemical reactions between ions of sodium, potassium, calcium, and chlorine.[98] For instance, neurologist A. D. Rockwell suggested in 1905 that the energy reserves and conductivity of neurons are crucial to mental health.[99] Rockwell thought neurons' ability to produce and conduct electricity played a central role in preventing mental disorders like hysteria and neurasthenia. These and other so-called "functional neuroses," Rockwell argued, result from "an impairment or interruption of the potential energy of cell life."[100]

According to this theory, squandering one's neuronal energy through sexual, nutritional, intellectual, or emotional excess could produce disastrous mental breakdowns.

## THE NEUROLOGICAL ROMANCE

In her romances, Corelli wove together these seemingly dissimilar ideas about neurasthenia, galvanism, cerebral localization theory, the voltaic cell, and the neuron as energy storehouse. The metaphor of the electric circuit was one common thread uniting these discourses. The concept of the electric battery – and the neuron as battery – likewise permeated these intellectual fields, and informed her novels' therapeutic mission. Corelli's readers literally "recharged their batteries" by reading uplifting tales of neurasthenics who overcome their despair to achieve spiritual transcendence and personal autonomy.

An examination of *A Romance of Two Worlds* demonstrates the centrality of the battery in Corelli's literary productions. In *Romance*, Corelli postulates that humans have electric organs in their bodies similar to those of electric fish, the biological models for the original Voltaic pile. These organs store energy in the body and allow people to project this energy onto their surroundings in the form of electric shocks and other manifestations. After reading about electric fish, the novel's heroine shares her new insight with Heliobas' electrically charged sister, Zara: "These fish have helped me to understand a great deal … [Heliobas] must have discovered the seed or commencement of electrical organs like those described, in the human body; he has cultivated them in you and in himself, and has brought them to a high state of perfection" (*RW* 197). Zara confirms the heroine's speculations: "As the muscles of the arm are developed by practice, so can the wonderful internal electrical apparatus of man be strengthened and enlarged by use" (*RW* 199).

Heliobas and Zara helpfully demonstrate the various uses of cultivated electric force, which include the ability to heal the sick, to generate protective electric force-fields surrounding the body, and to communicate telepathically with animals and divine beings. Zara uses electricity to protect herself from an aggressive suitor, Prince Ivan, who receives a massive electric shock when he tries to force himself upon her. Like electric fish, Heliobas and Zara can also use their internal electricity "as a healing power," as they demonstrate when they revive the dying Prince Ivan after his disastrous seduction attempt (*RW* 87). Similarly, mastery of internal electricity allows Heliobas and Zara to extend their lifespan and preserve the appearance of youth far longer than ordinary mortals. Zara, whose real age is thirty-eight, appears to be "seventeen, or at any rate not more than twenty" (*RW* 138). Heliobas uses electricity to communicate with his heavenly soul-mate, Azul, and also to "brain-electrify" his Saint Bernard, Leo, with whom he communicates telepathically (*RW* 174). Most impressively, Heliobas' electric powers help the protagonist embark on the life-changing out-of-body experience in which she views heaven itself, as described earlier in this chapter.

Corelli's musings about the manifold uses of human electrical organs creatively supplement Beard's conception of deficient nerve force by imagining the possibilities of nervous surplus. While Beard thought the highest aim of nervous energy was success in commerce and international relations, Corelli envisions spiritually transcendent ways to channel surplus nervous energy, including telepathy, telekinesis, and communication with divine forces. Since the most popular treatment for neurasthenic women was Silas Weir Mitchell's restrictive "rest cure," Corelli's active neurasthenic heroines seem iconoclastic, even ideologically radical. In contrast to the protagonist of Charlotte Perkins Gilman's "The Yellow Wallpaper" (1892), who slowly goes mad during a period of enforced confinement, Corelli's neurasthenic heroines traverse space and time with unprecedented rapidity. One suspects that Corelli's *Romance* empowered female patients of paternalistic physicians like Beard and Mitchell, or at least provided such women with an imaginative escape from these condescending male experts.

In later works, Corelli revised her theory to suggest that the brain served as the primary storehouse of human electricity, rather than a mysterious body part modeled after the organs of electric fish. Since Corelli equated electricity with light, heat, and God's love, her quasi-scientific theology became increasingly centered in the brain that stored these divine emanations. In *Ardath*, published three years after *Romance*, Corelli compares the brain to an "electric battery" that receives "electric messages of the Spirit."[101] Meanwhile, in *The Soul of Lilith*, Middle Eastern scientist El-Râmi-Zarânos describes how the energy flowing into and out of the brain can be conceived of as light, heat, or a divine force. "The vibrations of the human brain … like those emanating from the 'brain of heaven' are full of light and fire," he explains. These mystical vibrations, along with memories of past experiences, are "impressed in the curious little brain particles that lie in their various cells."[102]

Corelli's conception of the brain as battery may seem hopelessly outdated, as it is based on mid-Victorian cerebral localization theories and the neurasthenia diagnosis popular at the time. But a similar idea resurfaces in the 1999 film *The Matrix*, in which enslaved human brains channel the body's electric forces in order to power alien technology. The film, like Corelli's fiction, plays on the unsettling idea that the brain is interchangeable with an electric power source.[103] Whereas the film imagines tragic misuses of human brain power, however, Corelli aimed to show just how much people might accomplish by making the most of their energy reserves. Her fiction envisions miraculous feats performed on the strength of mental energy alone, offering readers inspiring vistas of health, youth, and endless vitality.

Over the course of her career, Corelli increasingly grounded her theology in the properties of the neuron itself. Corelli compared brain cells to tiny batteries in works like *Young Diana*. Diana willingly serves as a guinea pig for Dr. Dimitrius, who believes that radioactivity holds the secret to youth, health, and immortality. His scientific theories sound suspiciously like a combination of early-nineteenth-century vitalism and the neurasthenic logic described above: "Youth is in the cells of the brain. Should the cells become dry and withered it is because the soul has ceased to charge them with its energy. But when this is the case, it is possible – I say it is possible! – for science to step in. The spark can be re-energized, the cells can be re-charged."[104] Elsewhere in the novel, Corelli makes her comparison between cells and batteries more explicit. Dimitrius tells Diana, "For my purpose, you are not a woman, – you are simply an electric battery … our being is composed of millions of cells … the cells of the battery are dry – they must be recharged."[105] The vocabulary of this passage echoes that used by Beard and other physicians treating neurasthenia. But instead of electrotherapy, Diana's treatment involves immersion in radioactive fluid, which recharges her cells with electricity and invests her with everlasting youth and beauty. Corelli's unorthodox understanding of neuron doctrine leads her heroine not just to understand divinity – as in the author's earlier works – but to become divine.

Explicit comparisons between batteries and neurons are commonplace in biology and neurology textbooks today. But Corelli's linkage of these two structures in *The Young Diana* was based on her misunderstanding of their scientific properties. Corelli apparently did not know that batteries function by combining two differently charged chemicals to create an electric current. Both neurons and voltaic cells generate energy through chemical reactions of positively and negatively charged ions; Corelli, on the other hand, thought they stored electricity after the fashion of Leyden jars.[106] Her comparison of neurons and batteries would have been prescient, had she fully understood how batteries operate.

Corelli was certainly not alone in her misunderstanding of the neuron. Some *fin-de-siècle* neurologists, including Rockwell, likewise perceived the neuron as a reservoir of a limited quantity of nerve force. Freud, who began his career as a neurologist, expressed a similar view in his early work, *Project for a Scientific Psychology* (1895). In it, he describes a "neurone" filling with a certain quantity of energy, "while at other times it may be empty."[107] Although misguided, Rockwell and Freud were fairly typical of *fin-de-siècle* neurologists in their eagerness to borrow metaphors and concepts from the prestigious realm of physical science. Their idea that neurons "store" electricity in some mysterious fashion was an extension of the idea of conservation of energy into the realm of biology.

In fairness to Corelli, Rockwell, and Freud, relatively little was known about the electrical properties of neurons in the decades preceding the Second World War, leaving considerable room for speculation. Late-Victorian and Edwardian scientists agreed that neurons conducted electricity, but exactly how this occurred remained unclear until the 1950s, when new technologies like the electron microscope and the microelectrode revolutionized biology and neuroanatomy. Using these tools, mid-century scientists confirmed the existence of the synaptic gap, placing beyond doubt the validity of neuron doctrine. These scientists further demonstrated that both electrical stimuli and chemical reactions facilitate synaptic communication.[108]

Victorian and Edwardian scientists understood little about these processes, since existing microscopes were insufficiently powerful to observe the tiny gaps between neurons. Charles Sherrington, who coined the term "synapse" in 1897 to denote the hypothetical "surface of separation" between neurons, had difficulty imagining what exactly occurred at this important juncture. In 1906, he speculated: "Such a surface might restrain diffusion, bank up osmotic pressure, restrict the movement of ions, accumulate electric charges, support a double electric layer, alter in shape and surface tension with changes in difference of potential … or intervene as a membrane between dilute solutions of electrolytes of different concentration."[109] Sherrington's indecision is symptomatic of the early-twentieth-century controversy over whether synaptic transmission was primarily electric or chemical in nature, a dispute not resolved until the electronmicroscopic studies of the 1950s.[110] Similarly, the first accurate measurements of the speed of nerve impulses at the synapse were not carried out until 1922, a year after the publication of Corelli's last major work, *The Secret Power*.[111]

Mistaken though it was, Corelli's conception of the neuron as battery was a powerful spiritual stance. For Corelli, the neuron stored life energy and God's love, which could be utilized internally to preserve vitality or projected onto the outside world to perform miracles. One can imagine the appeal that this quasi-scientific principle must have had for world-weary Victorians and Edwardians, who were accustomed to finding religious doctrine incompatible with science. This concept of the neuron as energy storehouse especially appealed to women, who fantasized about converting their nervous energies into awe-inspiring mental powers. Corelli's romances promised healing not through a punitive "rest cure," but via restful immersion in a spiritual utopia that could recharge her audience's depleted nervous energies.

## RADIANT FAITH

Beginning in 1911 with the prologue to *The Life Everlasting*, Corelli began to consider new forms of energy that our brains might harness to achieve youth, health, and spiritual insight. In an effort to keep pace with the latest scientific discoveries, Corelli's later works substituted radium for electricity as the medium that constitutes the soul and communicates divine knowledge and love. Radium, the highly radioactive element discovered by Marie and Pierre Curie in 1898, was used in a variety of food and personal grooming products at the turn of the twentieth century because of its supposed health benefits. Since the dangers of radioactivity were not widely known until the 1920s, Corelli's revised Electric Creed presented in *The Life Everlasting* sounded sufficiently plausible to her many fans.[112]

Radium likely appealed to Corelli for two reasons: first, because it was discovered by a woman; and second, because she believed it was a self-renewing energy source. In later works like *Young Diana* and *The Secret Power* (1921), Corelli trumpeted her intense admiration for Marie Curie as a model of feminine scientific achievement. In *The Secret Power*, for instance, scientist heroine Morgana Royal compares herself to Marie Curie, "one of the greatest women of the age" who is nonetheless "allowed to remain in comparative obscurity" due to her gender.[113] Morgana invents an airship powered by radioactivity, to which she attributes divine significance: "Radio-activity is the chief secret of life. It is for us to learn how to absorb it into our systems as we grow, – to add by its means to our supplies of vitality and energy. It never gives out, – nor should we."[114]

The seemingly inexhaustible energy of radioactive materials gave Corelli a new way to explain her Electric Creed. She apparently did not know that radioactive materials lose energy over time, a fact established around 1900.[115] Like Corelli's earlier works, her later novels describe the brain as the storehouse of the energy emanating from the soul or from God. But Corelli now envisioned the "soul" consisting of "eternal radio-activity – capable of exhaustless energy" (*LE* 14). As the "wonderful 'discovery' of the immediate hour," radium intrigued Corelli due to its novelty, not to mention its ability to spontaneously produce light and heat (*LE* 18–19). To Corelli, these properties suggested the immortality of the soul and the resurrection of the dead: "Heat and Light are the composition of Life; – and the Life which this radio-activity of the Soul generates *in* itself and *of* itself, can never die" (*LE* 19). She continued, "It was … found that radium kept on producing heat *de novo* so as to keep itself always a fraction of a degree *above the surrounding temperature;* also that it spontaneously *produced electricity*. Does this teach no lesson on the resurrection of the dead?" (*LE* 20). For Corelli, radium's self-renewing qualities promised eternal conservation of energy within the human organism; any energy lost through nervous expenditure or ordinary wear and tear would be immediately replaced. Under these circumstances, eternal life seemed like a realistic possibility, one Corelli would later explore via the immortal heroine of *Young Diana*.

Corelli's ideas about radium seem less far-fetched when viewed alongside early-twentieth-century scientific writings on this topic. *Fin-de-siècle* scientists were amazed by the properties of newly discovered radioactive elements, and their prose conveys their sense of wonder. Marie Curie wrote in her 1904 article "Radium and Radioactivity" that the discovery of these phenomena "extends the limits of our knowledge" and forces us to recognize "how limited is our direct perception of the world which surrounds us, and how numerous and varied may be the phenomena which we pass without a suspicion of their existence." Curie confessed her amazement that these new elements emitted radiation "spontaneously and continuously."[116] A publication of the Roentgen Society of London described the discovery of radium still more exuberantly, calling the element "the most precious and magical powder ever dreamed of by alchemist or artist of romance":

> The wonder of this powder [radium], incessantly and without loss, under any and all conditions, pouring forth by virtue of its own intrinsic property powerful rays capable of penetrating opaque bodies … can perhaps be realized when

> we reflect that it is as marvellous as though we should dig up a stone which, without external influence or change, continually poured forth light or heat, manufacturing both in itself, and not only continuing to do so without appreciable loss or change, but necessarily having always done so for countless ages, whilst sunk beyond the ken of man in the bowels of the earth.[117]

Corelli would later use remarkably similar phrases to describe the marvels of radium, suggesting that she almost certainly read these or similar scientific sources.

In light of the above passages, Corelli's transcendent applications of radioactivity seem more plausible. In *The Life Everlasting*, Corelli describes the benefits of developing one's inner radioactivity in order to master the secrets of eternal life, youth, and love. The heroine's mentor and soul mate, Rafel Santoris, explains that commanding "the forces of nature" at work in one's body and surroundings results in "health, youth, and power! … Command of oneself! Command of body, command of spirit, and so on through an ever ascending scale!" (*LE* 237). Such "command" is achieved via the spirit or soul, which can "supply revivifying rays to every atom and cell in your body without stint or cessation. It is an exhaustless supply of 'radium' from which the forces of your life may draw perpetual sustenance" (*LE* 376). Santoris also uses his spiritual radioactivity to read thoughts, defend himself from enemies, and influence those who come into his own circle of electric attraction. By following a rigorous course of study at a secluded monastery, the heroine gains similarly miraculous powers.

Throughout *The Life Everlasting*, the heroine contrasts her own radioactive vitality to the depleted energies of her yachting companion, a neurasthenic heiress named Catherine Harland. Miss Harland is a self-proclaimed invalid who regularly receives electrotherapy from her incompetent physician, Dr. Brayle. The heroine disdains Miss Harland for her lack of willpower and sees her illness as a desperate ploy to gain sympathy and attention:

> In my opinion Catherine was robust enough, but it was evident that from a very early age she had been given her own way to the fullest extent, and had been so accustomed to have every little ailment exaggerated and made the most of that she had grown to believe health of body and mind as well-nigh impossible. (*LE* 59)

Dr. Brayle, too, comes in for his share of the heroine's contempt. She likens him to "an Italian poisoner of old time" and implies that he takes advantage of Miss Harland's credulity (*LE* 60).

Since the prologue to *The Life Everlasting* indicates that Corelli identified strongly with her protagonist, it seems likely that the author viewed the diagnosis and treatment of neurasthenia with skepticism. Her exposure of Dr. Brayle as a quack may well be an implied condemnation of Beard, Mitchell, and like-minded physicians. Nonetheless, Corelli's indebtedness to the logic of neurasthenia – itself an extension of the law of conservation of energy – is nowhere more evident than in this novel. Miss Harland's nervous deficiency serves as the background against which the heroine's transcendent nervous surplus gains legibility and (arguably) a certain degree of plausibility. Both characters operate according to a neurasthenic logic in which the energy reserves of the nervous system determine the individual's happiness and potential for meaningful action.

## CONCLUSION

Strange as Corelli's Electric Creed may seem to modern readers, her faith was grounded in the latest research in neuroscience and radioactivity, creatively adapted to fit late-Victorian spiritualist traditions. She was certainly not the only writer of her age to attribute spiritual significance to neurological ideas, though she was perhaps the most imaginative. Scientists themselves sought inspirational meanings in recent discoveries, as should be clear from the Romantic excitement surrounding electricity, the Edwardian enthusiasm for radioactivity, and various metaphysical interpretations of the first law of thermodynamics. The general excitement about neurons around the time Corelli wrote may likewise have influenced her beliefs. Scientific enthusiasm for this recently discovered but poorly understood unit of the nervous system is a palpable undertone running through the works of Cajal, Sherrington, and other late-Victorian and Edwardian neurologists, even if none of them imagined nerves and nervous energies as direct pathways of communication with God.

Corelli's most important function in the debates over neuron doctrine, however, was her role as a popularizer of scientific theories. As an opponent of scientific atheism who dabbled in spiritualism, Corelli was an unlikely but effective champion of neuron doctrine. This bestselling author helped bring the intensely controversial theory to the masses at a time when scientists themselves fiercely debated the details of neuron structure and function. Moreover, she made the edgy new concept palatable to laypeople by fusing it with her popular spiritual philosophy. The neuron doctrine as Corelli understood it had particular appeal for women, who sought uplifting alternatives to the punitive "rest cures" offered by their physicians.

From a literary perspective, Corelli deserves more credit for the success of her unusual blend of spirituality and science. In contrast to Stevenson or Wells, whose dystopian romances had antiscientific undertones, Corelli expanded and enriched the romance genre by putting a positive spin on neurological ideas. Corelli's inspired fusion of neurology and the romance helped readers reconcile spirituality with scientific materialism. Further, by envisioning the romance as therapy for a range of spiritual, emotional, and physical ills, Corelli heightened the seriousness of a genre that was sometimes unfairly dismissed as light, escapist reading and boys' adventure stories.

But because Corelli's fictions were based on misunderstandings of neurological innovations, one might justly question how far she advanced acceptance of cerebral localization. Corelli deliberately overlooked the materialist implications of neurological discoveries, such as the questionable role of the soul or will within a mechanistic understanding of brain function. Instead, Corelli promoted a transcendental vision of the neuron as storehouse and conductor of spiritual energy. This creative vision soothed her troubled readers, and temporarily restored harmony between physical and spiritual conceptions of mental function. But it succeeded primarily by capitalizing on areas of scientific ignorance or uncertainty. One could say practically anything about the neuron, and the newly discovered element radium, precisely because so little was known about them around the turn of the twentieth century. By the mid-twentieth century, increased scientific knowledge about neurons and radioactivity rendered Corelli's ideas obsolete. Corelli's brilliantly original "revision" of neurological concepts reminds us how difficult it was for Victorians to embrace all aspects of cerebral localization, particularly its biological determinism.

# *Epilogue: looking forward*

Though this study focuses on nineteenth-century scientific debates, the controversies I examine have not gone away. Fears spawned by cerebral localization and the related anxiety surrounding cerebral automatism have persisted in twentieth- and twenty-first-century art, literature, and popular culture. The most obvious examples come from science fiction, a genre directly descended from nineteenth-century Gothic novels like *Frankenstein* and *Jekyll and Hyde*.[1] In the last several decades, science fiction novels and films have addressed concerns surrounding cerebral localization, psychosurgery, and interfacing between brains and computers. The latter possibility seems to evoke the same fears as Victorian theories of cerebral automatism: namely, the idea that humans are mere machines, lacking free will and spiritual significance. Like the mechanistic vampire villain of Bram Stoker's *Dracula*, computers behave according to a system of rigidly stereotyped codes rather than responding to situations morally or empathically. As Oliver Sacks explains, computers lack the functions associated with the right hemisphere of the human brain, so that computational models of brain function are inadequate to explain the full range and complexity of personality, human relationships, and our sense of individual identity or "selfhood."[2]

Accordingly, much modern science fiction explores the ominous notion that human brains are (or are in the process of becoming) computers lacking any residue of "mind," "soul," or moral inhibition. For instance, in physician-litterateur Michael Crichton's *The Terminal Man* (1972), neurosurgeons implant computer-controlled electrodes in the brain of an epileptic man, Harold Benson, to help him suppress his violent seizures.[3] In one doctor's words, "this procedure represents the first direct link between a human brain and a computer."[4] But because the neurosurgeons place the electrodes too close to pleasure centers in the brain, Benson starts to provoke seizures on purpose and descends further into madness, becoming a homicidal maniac. Ironically, Benson is a computer programmer with delusional fears that computers are taking over the world. The tragic outcome of Benson's surgery seemingly validates his psychotic ruminations.

The 1980s science fiction subgenre of cyberpunk picks up where *The Terminal Man* leaves off, further speculating about the disastrous possibilities of brain–computer interfacing. As in Crichton's novel, such interfacing depends upon a sophisticated understanding of cerebral localization. William Gibson's *Neuromancer* (1984), for example, depicts a near future in which "open circulation between human and artificial intelligence" has become the norm.[5] In Gibson's fictional universe, psychosurgeons routinely implant microchips in human brains and nerves, giving patients superhuman capabilities. As a result, many future humans sport bionic limbs, brain–computer interfaces, and optical implants. Meanwhile, doctors utilize "vat-grown" flesh to repair or replace organs, including injured brain tissue.

If such sophisticated surgical techniques suggest the attainment of cerebral localization's highest goals, they also amply fulfill its Gothic promise. When surgical brain enhancement becomes routine, Gibson's novel suggests, ordinary human beings will be viewed as primitive "meat," a poor substitute for the sophisticated, emotionless intelligence of computers.[6] Indeed, the novel's absurd violence suggests that human life is scarcely valued at all. In a bleak future where humans can be cloned, cryogenically frozen, or surgically manipulated to resemble computers, essentialist categories like "humanity" and "nature" become radically destabilized.[7] Like the Gothic fiction that preceded it, then, cyberpunk often contains an antiscientific message, even as it invites us to marvel at the power and complexity of futuristic technologies.

Arguably, recent fictions of science have made the largest impact when translated onto the big screen. Blockbuster science fiction films of the last half-century have brought concerns about cerebral localization, mind control, and brain–computer interfacing to even broader audiences. In movies like *Invasion of the Body Snatchers* (1956) and *The Children of the Damned* (1960), invading alien life forms obliterate people's personalities, turning human beings into emotionless automata. As Susan Sontag perceptively notes, the human automaton is "the vampire fantasy in new dress … the danger is understood as residing in man's ability to be turned into a machine."[8] By implicitly comparing humans and machines, such films update late-Victorian concerns about cerebral automatism exemplified in *Dracula* and other works. *Invasion of the Body Snatchers* was remade three times (in 1978, 1993, and 2007), suggesting that its ominous message continues to haunt us.

More recent films likewise emphasize fear of mind control and explore the possibility of brains interfacing with malevolent computers. In Canadian filmmaker David Cronenberg's *eXistenZ* (1999), for instance, video games tap directly into players' neural networks, blurring the line between virtual reality and ordinary human consciousness. In the popular *Matrix* trilogy (1999, May 2003, November 2003), meanwhile, enslaved human brains channel the body's electric forces in order to power alien technology. Like Victorian novel-readers, modern moviegoers fear that "alien" scientific theories and technologies could rob them of dignity and personhood. These fears have only intensified as scientific disciplines like psychiatry, psychology, and neurology became more established social institutions over the course of the twentieth century.

In the twenty-first century, collaborations between bodies, brains, and machines are no longer restricted to science fiction. As early as 1991, Haraway's "Cyborg Manifesto" highlighted the degree of cooperation already extant between human bodies and machines, especially (but not exclusively) within the realm of medicine: for instance, injured or disabled individuals using machines to increase their mobility, or heart patients with pacemakers.[9] Almost two decades later, scientists can communicate with individual neurons and networks of neurons using nanotechnology. Chemist Charles Lieber, for instance, demonstrated in 2006 that "an array of nanowire transistors could be used to detect, stimulate, or dampen electrical signals from neurons," as well as to detect the release of the neurotransmitter dopamine. These mechanical interfaces allowed scientists to send and receive signals to and from individual neurons "in a way that … mimics the cells' own communication habits." While this technology does not yet have obvious medical applications, Lieber envisions a time when his experiments "could lead to more precise neural prostheses to treat blindness or neurological diseases."[10]

More dramatically, perhaps, the potential of new brain–computer interfacing (BCI) technology has recently gained public attention. For instance, a November 2008 episode of *60 Minutes* discussed how patients paralyzed by traumatic injuries and degenerative disorders (such as ALS) can communicate using BCI. Patients wear a cap that picks up the electrical activity of their brains and translates those electrical signals into letters that appear on a computer screen. Using this technology, paralyzed individuals can compose words and sentences that are then pronounced by the computer. Currently, this technology is rather slow, but its "life-changing" potential is already clear to scientists and patients. CBS reporter Scott Pelley suggests that BCI is "like a sudden leap in human evolution – a leap that could one day help paralyzed people walk again and amputees to move bionic limbs."[11]

Predictably, this new technology has aroused both optimism and suspicion from a public steeped in dystopian science fiction films. Pelley, for example, compares BCI to mind control or thought reading, echoing concerns voiced in cyberpunk fiction (where computers subliminally influence human thoughts through implanted microchips). Neuroscientist Jonathan Wolpaw reassures him that BCI "requires the cooperation of the [patient]" and that "it is certainly not mind control and it's different from reading people's thoughts."[12] The reporter's misunderstanding suggests that as in the Victorian period, popular fiction continues to shape public views of cerebral localization and new technologies that exploit our growing awareness of brain function.

So far, I have tried to demonstrate the continuity of ideas surrounding brain function and mental illness during the last fifty years, or even the last hundred and fifty years. But it is also essential to keep in mind the many significant ways in which mid-twentieth-century psychiatry differed from its present-day counterpart. For the greater part of the twentieth century, medical professionals relied mainly on talk therapy and various forms of depth psychology, particularly therapies derived from Freudian analysis. By the end of the twentieth century, however, the pendulum had begun to swing once more toward biological determinism.[13] Like late-Victorian scientists who attributed mental disturbances to bad heredity, cortical lesions, or to an imbalance of brain hemispheres, twenty-first-century physicians understand mental illnesses primarily as physiological disorders of the brain. In medical practice, this means that a person's own account of his or her experiences matters less to his or her physician than the presence of alleged "chemical imbalances" in the brain that can be most effectively treated with psychoactive medication.[14]

This clinical sea change will be reflected in the fifth edition of the American Psychiatric Association's *Diagnostic and Statistical Manual of Mental Disorders* (forthcoming in May 2013). According to a May 2009 article in the *Los Angeles Times*, the revised version of the *DSM* will be more "science based" than the fourth edition (1994): "Brain imaging and other technologies, plus new knowledge on biological and genetic causes of many disorders, have almost guaranteed significant alterations in how many mental afflictions are described."[15] Whether or not this biological turn in psychology represents a positive change for patients is debatable. In fact, some physicians fear that patients will be over-medicated as a result of more biologically oriented diagnoses.[16]

Nonetheless, the *DSM V* represents the culmination of two decades of neurological optimism within the scientific community. This excitement has long since trickled down to policymakers and even literary critics. The first President George Bush hailed the 1990s as "the decade of the brain," waxing optimistic about the potential of cutting-edge neuroscience to find cures for Alzheimer's, depression, epilepsy, AIDS, and even drug dependency.[17] Scientists heeded this clarion call by redoubling efforts to understand brain disease and the nature of cognition and emotion, using tools that David Ferrier never dreamed of. By employing the sophisticated, non-invasive brain mapping technology of functional magnetic resonance imaging (MRI) scanners, scientists have gained a more sophisticated understanding of which portions of the brain become active during specific behaviors and emotional responses.

Moreover, scientists have an improved understanding of what occurs in our brains at the microscopic level, within the neuron itself or even at the synapse, that most elusive juncture between neurons. Cutting-edge nanotechnology has allowed scientists to measure the electrical activity of neuronal networks, and even to communicate with individual neurons. Such technologies are still far too new to shed any light on higher-order brain functions at present.[18] But even so, some neurologists have already begun to speculate about the philosophical implications of such a precise understanding of neuronal function. Joseph LeDoux recently suggested, for instance, that human personality or "selfhood" can be best understood at the synaptic level, arguing that "your 'self,' the essence of who you are, reflects patterns of interconnectivity between neurons in your brain."[19] He continues, "for many people, the brain and the self are quite different. I hope to show … that this is not the case."[20] While such a conclusion might seem like biological reductionism at its most extreme, LeDoux points out that experience as well as genetics plays a role in shaping synaptic transmission and brain function more generally.

Most recently, literary critics have jumped on the cognitive science bandwagon, using knowledge about the structure and functions of the brain to arrive at "objective" truths about artistic works. As Alan Richardson and Francis F. Steen observe, the emergent field "broadly defined as cognitive literary criticism" draws its inspiration and methodology from various fields within cognitive science, including "artificial intelligence, cognitive psychology, post-Chomskian linguistics, philosophy of mind, neuroscience, and evolutionary biology."[21] The goal of blending humanities and cognitive neuroscience, Mark Turner explains, is to create "a practical, sustainable, intelligible, intellectually coherent paradigm for answering basic and recurring questions about the cognitive instruments of art, language, and literature."[22]

Cognitive literary criticism overlaps with another emergent movement known as literary Darwinism, whose practitioners apply Darwinian evolutionary theory and sociobiological theories to literary texts, drawing frequently upon the work of renowned biologist Edward O. Wilson.[23] According to proponents, cognitive literary criticism and literary Darwinism potentially hold the answers to big questions such as why literature evolved in the first place and what its biological usefulness might be.[24] These emergent schools of criticism have already produced many novel insights and encouraged pathbreaking interdisciplinary thinking. While the best work of this type avoids anachronism by acknowledging scientific paradigm shifts, some literary Darwinists and cognitive literary critics naively assume that the latest scientific ideas are objective, timeless "truths." Looking over the discarded scientific hypotheses of previous centuries, one can easily see the dangers inherent in such assumptions.

Despite this widespread climate of neurological enthusiasm, scientific attempts to ground human behavior in cerebral structure and brain chemistry still arouse considerable anxiety. For example, in the past few decades, the introduction of powerful new medications to treat depression, bipolar disorder, attention deficit disorder, and other conditions provoked backlash from some doctors and patients, who resented the implication that human behavior could be fully explained by neurochemistry.[25] Psychiatrist Arthur Kleinman suggests, for instance, that "we are in danger as a society of flattening out our moral life" by overprescribing drugs that make it difficult for people to engage with suffering in a meaningful way.[26] Similarly, while MRI brain-scan techniques and BCI offer remarkable diagnostic and therapeutic benefits, a prominent historian of science countered in 1990 that "the progressive reductionism of mind to physiology" has "impoverished conceptions of mental life."[27] Apparently, the specter of biological determinism is no less disturbing to us than it was to our Victorian predecessors. Perhaps this helps to explain why novels like *Jekyll and Hyde*, *Dracula*, and *The War of the Worlds* remain enormously popular and culturally influential.

# *Notes*

## INTRODUCTION: CEREBRAL LOCALIZATION AND THE LATE-VICTORIAN GOTHIC ROMANCE

1 Bram Stoker, *Dracula*, ed. Nina Auerbach and David Skal (New York: W. W. Norton and Company, 1997), 71.
2 Oliver Sacks, *The Man Who Mistook His Wife for a Hat and Other Clinical Tales* (New York: Simon & Schuster, 1998), 3.
3 Stanley Finger, *Minds Behind the Brain: A History of the Pioneers and Their Discoveries* (New York: Oxford University Press, 2000), 150.
4 Finger, *Minds Behind the Brain*, 161–2.
5 Laura Otis, "Howled out of the Country: Wilkie Collins and H. G. Wells Retry David Ferrier," in Anne Stiles, ed., *Neurology and Literature, 1860–1920* (New York: Palgrave, 2007), 27–51, 29.
6 Otis, "Howled out of the Country," 30–1.
7 George Levine, *The Realistic Imagination: English Fiction from Frankenstein to Lady Chatterley* (University of Chicago Press, 1981), 18.
8 Levine, *The Realistic Imagination*, 11.
9 Lawrence Rothfield, *Vital Signs: Medical Realism in Nineteenth-Century Fiction* (Princeton University Press, 1992), 120–1.
10 Jeremy Tambling fleshes out this analogy between realist narrator and physician in "*Middlemarch*, Realism, and the Birth of the Clinic," *ELH* 57.4 (Winter 1990): 939–60, 947–8.
11 Émile Zola, "The Experimental Novel," in *The Experimental Novel and Other Essays*, trans. Belle Sherman (New York: Cassell Publishing, 1893), 25.
12 Zola, "The Experimental Novel," 25.
13 Rick Rylance, *Victorian Psychology and British Culture, 1850–1880* (New York: Oxford University Press, 2000), 3.
14 Nicholas Dames, *The Physiology of the Novel: Reading, Neural Science, and the Form of Victorian Fiction* (New York: Oxford University Press, 2007), 48.
15 For an overview of recent scholarship on Victorian psychology and literature, including a more thorough exploration of its bias in favor of high realism and George Eliot, see Anne Stiles, "Victorian Psychology and the Novel," *Literature Compass* 5.3 (2008): 668–80.

16 See Julia Reid, *Robert Louis Stevenson, Science, and the* Fin de Siècle (New York: Palgrave Macmillan, 2006), 1–53; Kelly Hurley, *The Gothic Body: Sexuality, Materialism, and Degeneration at the* Fin de Siècle (New York: Cambridge University Press, 1996), especially pp. 3–20; and Robert Mighall, *A Geography of Victorian Gothic Fiction: Mapping History's Nightmares* (New York: Oxford University Press, 1999), chapters 4 through 6.

17 Nicholas Dames, *Amnesiac Selves: Nostalgia, Forgetting, and British Fiction, 1810–1870* (New York: Oxford University Press, 2001), 167–205; Jenny Bourne Taylor, *In the Secret Theatre of Home: Wilkie Collins, Sensation Narrative, and Nineteenth-Century Psychology* (New York: Routledge, 1988).

18 Stoker's biographers disagree about what subjects he studied at Trinity College and when he graduated. The controversial details of Stoker's educational background will be discussed at length in Chapter 2.

19 Shortly after the publication of *Jekyll and Hyde*, Myers's article "Multiplex Personality" appeared in *The Nineteenth Century* 20 (November 1886): 648–66. Stanley Finger mentions Stevenson's potential influence on Lewis Bruce in *Origins of Neuroscience: A History of Explorations into Brain Function* (New York: Oxford University Press, 1994), 396–7.

20 The word "pseudoscience" is used here with some reservations. Recent historical and literary criticism makes clear that many scientific discourses now regarded as illegitimate (such as mesmerism, phrenology, and psychical research) at one time garnered widespread support among medical practitioners. See, for example, Helen Small, "Introduction" to *The Lifted Veil and Brother Jacob* (New York: Oxford University Press, 1999), ix–xxxviii, xviii n.20.

21 Ian Duncan, *Modern Romance and Transformations of the Novel: The Gothic, Scott, Dickens* (New York: Cambridge University Press, 1992), 10.

22 This list of formal and thematic elements of Gothic fiction is by no means exhaustive. For more complete catalogues of Gothic horrors, see the introductions to David Punter, *The Literature of Terror: A History of Gothic Fiction from 1765 to the Present Day* (London: Longman, 1980) and Eugenia DeLamotte, *Perils of the Night: A Feminist Study of Nineteenth-Century Gothic* (New York: Oxford University Press, 1990). My brief overview of early Gothic fiction draws from these sources, and from Heather Wozniak's excellent Ph.D. dissertation, "Brilliant Gloom: The Contradictions of British Gothic Drama, 1768–1823" (University of California, Los Angeles, 2008).

23 Duncan, for instance, describes romance as "modernity's vision of worlds it has superseded, charged with a magic of estrangement, perils, and loss." *Modern Romance*, 9.

24 Duncan, *Modern Romance*, 13.

25 Nicholas Daly, *Modernism, Romance, and the* Fin de Siècle: *Popular Fiction and British Culture, 1880–1914* (New York: Cambridge University Press, 1999), 18–19, 23.

26 Peter Garrett, *Gothic Reflections: Narrative Force in Nineteenth-Century Fiction* (Ithaca: Cornell University Press, 2003), 3.

27 Fred Botting, *Gothic Romanced: Consumption, Gender, and Technology in Contemporary Fictions* (New York: Routledge, 2008), 9.
28 Daly, *Modernism, Romance, and the* Fin de Siècle, 12–16.
29 Robert Mighall considers the Gothic a literary mode with a particular relationship to history, rather than a set of narrative conventions per se (*A Geography of Victorian Gothic Fiction*, xix). He thus sees no difficulty in including late-Victorian works under the Gothic umbrella, including those like *Jekyll and Hyde*, in which urban settings replace the traditional moldy castles of eighteenth-century Gothic romances. Patrick Brantlinger, meanwhile, describes a category of *fin-de-siècle* fiction he calls "imperial Gothic," whose principal themes include "going native; an invasion of civilization by the forces of barbarism; and the diminution of opportunities for adventure and heroism in the modern world." *Rule of Darkness: British Literature and Imperialism, 1830–1914* (Ithaca, NY: Cornell University Press, 1988), 229–30. Kelly Hurley argues that late-Victorian Gothic fiction is preoccupied with what she calls the ab-human subject, whose body is "characterized by its morphic variability, continually in danger of becoming not-itself, becoming other" (*The Gothic Body*, 3–4).
30 See chapters 4 through 6 in Alice Jenkins, *Space and the "March of Mind": Literature and the Physical Sciences in Britain 1815–1850* (New York: Oxford University Press, 2007).
31 DeLamotte, *Perils of the Night*, 21–3.
32 Steven Bruhm, *Gothic Bodies: The Politics of Pain in Romantic Fiction* (Philadelphia: University of Pennsylvania Press, 1994), xiii–xxii.
33 Mighall, *A Geography of Victorian Gothic Fiction*, xxiv.
34 William Greenslade, *Degeneration, Culture, and the Novel 1880–1940* (Cambridge University Press, 1994), 33–4.
35 Hurley, *The Gothic Body*, 11; Stephen Heath, "Psychopathia sexualis: Stevenson's *Strange Case*," *Critical Quarterly* 28.1 (1986), 93–108, 103.
36 Hurley, *The Gothic Body*, 6.
37 Jean Pierre Marie Flourens, *Phrenology Examined* (1846), paraphrased in Anne Harrington, *Medicine, Mind, and the Double Brain: A Study in Nineteenth-Century Thought* (Princeton University Press, 1987), 9.
38 William Benjamin Carpenter, "On the Doctrine of Human Automatism," *Contemporary Review* 25 (1875): 397–416, 397.
39 Carpenter, "On the Doctrine of Human Automatism": 403.
40 The historical continuity between phrenology and cerebral localization theories is the premise of Robert Young's *Mind, Brain and Adaptation in the Nineteenth-Century: Cerebral Localization and Its Biological Context from Gall to Ferrier* (New York: Oxford University Press, 1990). Young suggests that Gall was asking the right questions about brain function, but coming up with the wrong answers since he used intuitive rather than scientific methods. Later researches like Ferrier, Hitzig, and Jackson utilized more reliable experimental methods to find more accurate answers to the same basic questions about the brain (4, 9–53).

41 Young, *Mind, Brain and Adaptation*, 3, 9.
42 Jenny Bourne Taylor and Sally Shuttleworth, eds., *Embodied Selves: An Anthology of Psychological Texts, 1830–1890* (New York: Oxford University Press, 1998), 4–5.
43 Harrington, *Medicine, Mind, and the Double Brain*, 9n.
44 Alan Richardson, *British Romanticism and the Science of the Mind* (Cambridge University Press, 2001), 12.
45 I am indebted to Alan Richardson for this observation.
46 Edward Hungerford, "Poe and Phrenology," *American Literature* 2.3 (November 1930): 209–31.
47 On phrenology in Brontë's fiction, see Dames, *Amnesiac Selves*, 76–124. On Eliot and phrenology, see B. M. Gray, "Pseudoscience and George Eliot's 'The Lifted Veil,'" *Nineteenth-Century Fiction* 36.4 (1982): 407–23.
48 Finger, *Origins of Neuroscience*, 41, 438–9.
49 Frances Power Cobbe, "Light in Dark Places," in *The Modern Rack: Papers on Vivisection* (London: Swan Sonnenschein, 1889), 210.
50 Otis, "Howled out of the Country," 31.
51 See Otis, "Howled out of the Country," 37–47.
52 These debates originally took place in correspondence and essays written in the 1880s. Some of this material has since been republished in Janet Adam Smith, ed., *Henry James and Robert Louis Stevenson: A Record of Friendship and Criticism* (London: R. Hart Davis, 1948).
53 William James, "Frederic Myers's Service to Psychology" (1901), in *Essays in Psychical Research* (Cambridge, MA: Harvard University Press, 1986), 201.
54 William James, "Frederic Myers's Service to Psychology," 193–4.
55 I borrow the term "classical neurology" from Luria via Sacks, *The Man Who Mistook His Wife for a Hat*, 234.
56 Sacks, *The Man Who Mistook His Wife for a Hat*, 4–5.
57 Sacks, *The Man Who Mistook His Wife for a Hat*, 4–5.
58 A. R. Luria, *The Mind of a Mnemonist*, trans. Lynn Solotaroff (New York: Basic Books, 1968), 159.
59 Michel Foucault memorably describes the confessional bent of nineteenth-century sexology in *The History of Sexuality, Volume One: An Introduction*, trans. Robert Hurley (New York: Vintage Books, 1978). See especially part three, "Scientia Sexualis."
60 On the objectivity of nineteenth-century case reports, see Michel Foucault, *The Birth of the Clinic: An Archaeology of Medical Perception*, trans. A. M. Sheridan Smith (New York: Vintage Books, 1994), especially chapter 7, "Seeing and Knowing." See also Jan Goldstein, "The Uses of Male Hysteria: Medical and Literary Discourse in Nineteenth-Century France," *Representations* 34 (1991): 134–65, 138.
61 Lilian Furst, *Between Doctors and Patients: The Changing Balance of Power* (Charlottesville: University Press of Virginia, 1998), 6.
62 David Ferrier, "The Brain of a Criminal Lunatic," *Brain: A Journal of Neurology* 5 (April 1882): 62–73, 62.

63 Ferrier, "The Brain of a Criminal Lunatic": 71.
64 Roslynn D. Haynes, *From Faust to Strangelove: Representations of the Scientist in Western Literature* (Baltimore: Johns Hopkins University Press, 1994), 17.
65 Brian Aldiss, *Billion Year Spree: The True History of Science Fiction* (New York: Doubleday, 1973), 18.
66 Patrick Brantlinger, "The Gothic Origins of Science Fiction," *NOVEL: A Forum on Fiction* 14.1 (1980): 30–43, 31.
67 Daly, *Modernism, Romance, and the* Fin de Siècle, 8.
68 Daly, *Modernism, Romance, and the* Fin de Siècle, 16; Levine, *The Realistic Imagination*, 9.
69 Margaret Atwood, Introduction to H. G. Wells, *The Island of Doctor Moreau*, ed. Patrick Parrinder (New York: Penguin, 2005), xiii–xxvii, xviii–xix.
70 Some of the best treatments of nineteenth-century romances include Daly, *Modernism, Romance, and the* Fin de Siècle; Reid, *Robert Louis Stevenson, Science, and the* Fin de Siècle; and Duncan, *Modern Romance*.
71 Daly, *Modernism, Romance, and the* Fin de Siècle, 27.
72 Reid, *Robert Louis Stevenson, Science, and the* Fin de Siècle, 32.
73 Robert Louis Stevenson, "Pastoral" (1887), in *Robert Louis Stevenson: The Scottish Stories and Essays*, ed. Kenneth Gelder (Edinburgh University Press, 1989), 242–8, 247.
74 Stevenson, "Pastoral," 247.
75 Regarding the nervous impact of sensation fiction, see Taylor, *In the Secret Theatre of Home*, especially the introduction; and Dames, *Amnesiac Selves*, 76–102.
76 "The Sensation Times," *Punch* 44 (May 1863): 193, quoted in Matthew Sweet, "Introduction" to Wilkie Collins, *The Woman in White* (New York: Penguin, 1999), xviii.
77 Robert Louis Stevenson, "A Gossip on Romance," in Glenda Norquay, ed., *R. L. Stevenson on Fiction: An Anthology of Literary and Critical Essays* (Edinburgh University Press, 1999), 51–64, 55.
78 Robert Louis Stevenson, "A Humble Remonstrance," in Smith, *Henry James and Robert Louis Stevenson*, 86–100, 95.
79 Stevenson, "A Humble Remonstrance," 90.
80 Stevenson, "A Gossip on Romance," 52.
81 "A Chat With the Author of *The Time Machine*, Mr. H. G. Wells," quoted in Steven McLean, *The Early Fiction of H. G. Wells: Fantasies of Science* (New York: Palgrave, 2009), 2.
82 Wells referred to his early works as "scientific romances," but he was not the first to use the term, which originated with the lesser-known writer Charles Howard Hinton (Atwood, "Introduction," xvii).
83 H. G. Wells, "Introduction" to *The Scientific Romances of H. G. Wells* (London: Victor Gollancz, 1933), viii.
84 Wells, "Introduction" to *The Scientific Romances of H. G. Wells*, viii (emphasis in original).

85 Daniel Pick, *Faces of Degeneration: A European Disorder, c.1848–1918* (Cambridge University Press, 1989), 100–1.
86 Marie Corelli to George Bentley, November 13, 1886, Corelli Collection (Cat. No. GEN MSS 332), Yale University, Beinecke Rare Book and Manuscript Library.
87 Corelli to Bentley, November 15, 1886, Corelli Collection.
88 Henry James, "The Art of Fiction," in Smith, *Henry James and Robert Louis Stevenson*, 53–80, 56.
89 Many of Linda Hutcheon's works explore the concept of "complicitous critique," which she explains in the first chapter of *The Politics of Postmodernism*, 2nd edn. (New York: Routledge, 2002), 1–28.

## 1 ROBERT LOUIS STEVENSON'S *JEKYLL AND HYDE* AND THE DOUBLE BRAIN

1 Oscar Wilde, "The Decay of Lying: An Observation," in *The Soul of Man under Socialism and Selected Critical Prose*, ed. Linda Dowling (New York: Penguin, 2001), 167.
2 Fanny Osborne Stevenson, "Prefatory Note" to Robert Louis Stevenson, *The Strange Case of Dr. Jekyll and Mr. Hyde; Fables, Other Stories and Fragments* (London: Heinemann, 1924), xv–xvi. This version was reprinted from a 1905 preface to Stevenson's collected works published by Charles Scribner's Sons.
3 "Mr. R. L. Stevenson on Reading and Literature: An Interview," *The Argus*, April 11, 1893. Quoted in Richard Dury, "Crossing the Bounds of Single Identity: *Dr. Jekyll and Mr. Hyde* and a Paper in a French Scientific Journal," in Richard Ambrosini and Richard Dury, eds., *Robert Louis Stevenson, Writer of Boundaries* (Madison: University of Wisconsin Press, 2006), 237–51, 237.
4 Dury explains, "Later in the Preface… [Fanny] claims that the tale was written in three days and rewritten in another three, while Stevenson tells us that it was written over a space of about six weeks." "Crossing the Bounds of Single Identity," 239.
5 Myers's correspondence with Stevenson is reprinted in Paul Maixner, ed., *Robert Louis Stevenson: The Critical Heritage* (Boston: Routledge, 1981), 212–21.
6 Dury, "Crossing the Bounds of Single Identity," 248.
7 Jean-Pierre Naugrette, "Le texte et son double: le cas de M.P. et du Dr. Forsyth," *Otrante* 8 (1995–6): 149–59, and "L'Étrange case du doutor Pereira et du docteur Cardoso: essai sur la fonction cognitive et politique d'un mythe littéraire," *La Licorne* 55 (2000): 277–92. See also Jacqueline Carroy, *Hypnose, suggestion et psychologie: L'invention des sujets* (Paris: Presses Universitaires de France, 1991), 103–9, "Entre mémoire et oubli: les deux vies de Félida," *Revue internationale de psychopathologie* 5 (1992): 73–80, and "Le savant et ses doubles: Docteurs Jekyll et Misters Hyde du siècle dernier," *Frénesie* 10 (Spring 1992): 125–42.

8 Ian Hacking, "*Automatisme Ambulatoire*: Fugue, Hysteria, and Gender at the Turn of the Century," *Modernism/Modernity* 3.2 (April 1996): 31–43, 38.
9 Ian Hacking, *Rewriting the Soul: Multiple Personality and the Sciences of Memory* (Princeton University Press, 1995), 160.
10 Hacking, *Rewriting the Soul*, 162.
11 Hacking, "*Automatisme Ambulatoire*": 38.
12 In 1886, for instance, Myers wrote that "Félida's name at least is probably familiar to most of my readers." Frederic W. H. Myers, "Multiplex Personality," *The Nineteenth Century* 20 (1886): 648–66, 648.
13 Ernest Mesnet, *De L'Automatisme de la Mémoire et du Souvenir dans le Somnambulisme Pathologique* (Paris: Imprimerie de Félix Malteste, 1874).
14 Jenny Bourne Taylor and Sally Shuttleworth, eds., *Embodied Selves: An Anthology of Psychological Texts, 1830–1890* (New York: Oxford University Press, 1998), 96.
15 Julia Reid, *Robert Louis Stevenson, Science, and the* Fin de Siècle (New York: Palgrave, 2006), 96.
16 Stanley Finger, *Minds Behind the Brain: A History of the Pioneers and Their Discoveries* (New York: Oxford University Press, 2000), 151. See also Stanley Finger *et al.*, "'Dual Brain Action': The Case Studies of Lewis C. Bruce in the 1890s," *History of Psychiatry* 12 (2001): 59–71.
17 Regarding James Sully's friendship with Stevenson, see Ed Block, Jr., "James Sully, Evolutionary Psychology, and Late-Victorian Gothic Fiction," *Victorian Studies* 25.4 (1982): 443–67. On Stevenson's correspondence with Pierre Janet, see Hacking, *Rewriting the Soul*, 278n.
18 Robert Mighall, *A Geography of Victorian Gothic Fiction: Mapping History's Nightmares* (New York: Oxford University Press, 1999), 138; Patrick Brantlinger, *Rule of Darkness: British Literature and Imperialism, 1830–1914* (Ithaca, NY: Cornell University Press, 1988), 245.
19 Robert Louis Stevenson, *Strange Case of Dr. Jekyll and Mr. Hyde*, ed. Katherine Linehan (New York: W. W. Norton and Company, 2003), 48. Further references to this source will be cited parenthetically within the text.
20 Hacking, *Rewriting the Soul*, 172–3.
21 Hacking, *Rewriting the Soul*, 171.
22 See Elizabeth Loentz, *Let Me Continue to Speak the Truth: Bertha Pappenheim as Author and Activist* (Cincinnati, OH: Hebrew Union College Press, 2007), 210, 210n.
23 Hacking, *Rewriting the Soul*, 4.
24 Anne Harrington, *Medicine, Mind, and the Double Brain: A Study in Nineteenth-Century Thought* (Princeton University Press, 1987), 136. I am very much indebted to Harrington's work throughout this section of the chapter. Since she has written the definitive scholarly text on the subject of the double brain, my reliance on her work is to some extent inevitable, but it also attests to my deep appreciation of her intellectual contributions to the field.

25 Finger, *Minds Behind the Brain*, 150.
26 Elaine Showalter, "Dr. Jekyll's Closet," in *The Haunted Mind: The Supernatural in Victorian Literature* (London: Scarecrow Press, 1999), 75.
27 Showalter, "Dr. Jekyll's Closet," 75.
28 James Strachey writes that in Freud's early works such as *Project for a Scientific Psychology* (1895), Freud "intended to make it possible to describe mental events in purely neurological terms. This attempt culminated in the *Project* and was not long afterwards abandoned." Editor's Introduction to Josef Breuer and Sigmund Freud, *Studies on Hysteria*, trans. and ed. James Strachey (New York: Basic Books, 1957), xxiv.
29 Freudian readings of Stevenson's work are so numerous that it is impossible to include a complete list here. Some notable examples include Showalter, "Dr. Jekyll's Closet"; William Veeder, "Children of the Night: Stevenson and Patriarchy," in William Veeder and Gordon Hirsch, eds., *Dr. Jekyll and Mr. Hyde After One Hundred Years* (University of Chicago Press, 1988), 107–60; and Hilary Beattie, "Father and Son: The Origins of *Strange Case of Dr. Jekyll and Mr. Hyde*," *The Psychoanalytic Study of the Child* 56 (2001): 317–60.
30 Frederic W. H. Myers, "Multiplex Personality," quoted in Stevenson, *Strange Case of Dr. Jekyll and Mr. Hyde*, ed. Linehan, 135.
31 Reid, *Robert Louis Stevenson, Science, and the* Fin de Siècle, 27.
32 Martin Danahay suggests that "while Myers entitles his essay 'Multiplex personality,' he is actually interested … in the 'double' brain and examines cases that seem to reveal the duality of identity." But Myers's decision to examine Louis Vivet, the first recognized case of multiple personality, would seem to belie this argument. Danahay, "Introduction" to Myers, "Multiplex Personality," in Robert Louis Stevenson, *The Strange Case of Dr. Jekyll and Mr. Hyde*, 2nd edn., ed. Martin Danahay (Peterborough, Ontario: Broadview, 2005), 198.
33 Robert M. Young, *Mind, Brain and Adaptation in the Nineteenth Century: Cerebral Localization and Its Biological Context from Gall to Ferrier* (New York: Oxford University Press, 1990), 9–53.
34 Harrington, *Medicine, Mind, and the Double Brain*, 15.
35 Harrington, *Medicine, Mind, and the Double Brain*, 11.
36 F. A. Lange, quoted in Harrington, *Medicine, Mind, and the Double Brain*, 9.
37 Harrington, *Medicine, Mind, and the Double Brain*, 9n.
38 Harrington, *Medicine, Mind, and the Double Brain*, 10.
39 Young, *Mind, Brain and Adaptation*, 3.
40 Arthur Ladbroke Wigan, *A New View of Insanity: The Duality of the Mind* (1844), quoted in Taylor and Shuttleworth, *Embodied Selves*, 125.
41 Harrington, *Medicine, Mind, and the Double Brain*, 27.
42 Wigan, *A New View of Insanity*, quoted in Taylor and Shuttleworth, *Embodied Selves*, 127.
43 Steven J. Gould, *The Mismeasure of Man* (New York: W. W. Norton and Company, 1981), 74.

44 Harrington, *Medicine, Mind, and the Double Brain*, 71.
45 Harrington, *Medicine, Mind, and the Double Brain*, 73.
46 Gould, *The Mismeasure of Man*, 73–112.
47 Harrington, *Medicine, Mind, and the Double Brain*, 80.
48 Robert Louis Stevenson, "A Chapter on Dreams" (1888), quoted in Stevenson, *Strange Case of Dr. Jekyll and Mr. Hyde*, ed. Linehan, 91.
49 Oliver Sacks, *The Man Who Mistook His Wife for a Hat and Other Clinical Tales* (New York: Simon & Schuster, 1998), 4.
50 Sacks, *The Man Who Mistook His Wife for a Hat*, 4.
51 Harrington, *Medicine, Mind, and the Double Brain*, 79.
52 Sacks, *The Man Who Mistook His Wife for a Hat*, 4.
53 There have been additional changes in our understanding of hemispheric functioning since the Victorian era, though not as many as one might expect. Scientists now think that women are predominantly left-brained, and that linguistic functions and mathematical reasoning may involve both hemispheres to some extent, rather than only the left hemisphere. For instance, some left-handed individuals have speech localized in the right hemisphere, possibly due to problems in brain development that forced the right hemisphere to compensate for the left. See Sally Springer and Georg Deutsch, *Left Brain/Right Brain: Perspectives from Cognitive Neuroscience*, 5th edn. (New York: W. H. Freeman, 1997), especially chapter 3; and Finger, *Minds Behind the Brain*, 148.
54 Richard Proctor, "Have We Two Brains?" *The Cornhill Magazine* 31 (1875): 149–66, 157.
55 Corpus callosotomy, also called callosal section, is a surgical procedure used to treat severe epileptic seizures since the mid twentieth century.
56 Robert Louis Stevenson, "My First Book" (1894), in *Treasure Island*, ed. Emma Letley (New York: Oxford University Press, 1985), 192–200, 195.
57 G. Stanley Hall and E. M. Hartwell, "Bilateral Asymmetry of Function," *Mind* 9.33 (January 1884): 94–109, 102.
58 Harrington, *Medicine, Mind, and the Double Brain*, 89.
59 David Ferrier, "The Brain of a Criminal Lunatic," *Brain: A Journal of Neurology* 5 (April 1882): 62–73, 64, 66.
60 Finger, *Minds Behind the Brain*, 153.
61 One exception to this rule is Broca, who "warned others not to assume that the intellectual side of the brain can be determined with confidence just by knowing handedness." Finger, *Minds Behind the Brain*, 149.
62 Emphasis in original. Frederic W. H. Myers to Robert Louis Stevenson, February 27, 1886, quoted in Paul Maixner, *Robert Louis Stevenson*, 215.
63 Robert Macnish described the dual personalities of Mary Reynolds in *The Philosophy of Sleep* (1830). Reynold's case "became one of the most famous instances of double or divided consciousness during the nineteenth century." Taylor and Shuttleworth, *Embodied Selves*, 123.
64 Peter Morton, *"The Busiest Man in England": Grant Allen and the Writing Trade, 1875–1900* (New York: Palgrave, 2005), 105.

65 On Proctor's career, see Bernard Lightman, *Victorian Popularizers of Science, Designing Nature for New Audiences* (University of Chicago Press, 2007), 295–352; and Morton, *"The Busiest Man in England"*, 76, 78, 105.
66 Richard Proctor, "Dual Consciousness," *The Cornhill Magazine* 35 (January 1877): 86–105, 90.
67 Azam, quoted (in translation) in Proctor, "Dual Consciousness": 90.
68 Hacking, *Rewriting the Soul*, 175.
69 Proctor, "Dual Consciousness": 91.
70 Eugène Azam, *Hypnotisme, Double Conscience, et Altérations de la Personnalité* (Paris: Librairie J. B. Baillière et Fils, 1887), 73. Translations are my own, unless otherwise noted.
71 Gerard Manley Hopkins to Robert Bridges, October 28, 1886, quoted in Maixner, *Robert Louis Stevenson*, 228.
72 Stephen Heath, "Psychopathia Sexualis: Stevenson's *Strange Case*," *Critical Quarterly* 28.1 (1986): 93–108, 93.
73 Proctor, "Dual Consciousness": 90.
74 Proctor, "Have We Two Brains?": 155.
75 Dury, "Crossing the Bounds of Single Identity," 244.
76 Frederic W. H. Myers to Robert Louis Stevenson, February 27, 1886, quoted in Maixner, *Robert Louis Stevenson*, 216.
77 Proctor, "Have We Two Brains?": 155.
78 Proctor, "Have We Two Brains?": 155.
79 Proctor, "Have We Two Brains?": 156.
80 Hacking, *Rewriting the Soul*, 5.
81 On male hysteria in the nineteenth century, see Mark Micale, *Hysterical Men: The Hidden History of Male Nervous Illness* (Cambridge, MA: Harvard University Press, 2008).
82 Mark Micale, "Charcot and the Idea of Hysteria in the Male: Gender, Mental Science, and Medical Diagnosis in Late Nineteenth-Century France," *Medical History* 34 (1990): 363–411, 393.
83 See, for instance, Katherine Linehan, "Introduction" to Stevenson, *Strange Case of Dr. Jekyll and Mr. Hyde*, ed. Linehan, xi–xv, xiv; Martin Danahay, "Introduction" to Stevenson, *The Strange Case of Dr. Jekyll and Mr. Hyde*, 2nd edn., ed. Danahay, 11–25, 14; and Masao Miyoshi, "Dr. Jekyll and the Emergence of Mr. Hyde," *College English* 27.6 (1966): 470–80, 474.
84 Mighall, *A Geography of Victorian Gothic Fiction*, xxiv.
85 Robert Louis Stevenson, "A Humble Remonstrance," in Glenda Norquay, ed., *R. L. Stevenson on Fiction: An Anthology of Literary and Critical Essays* (Edinburgh University Press, 1999), 83.
86 Stevenson, "A Humble Remonstrance," 83.
87 Robert Louis Stevenson, "A Note on Realism," in *R. L. Stevenson on Fiction*, 71; Robert Louis Stevenson, "A Gossip on Romance," in *R. L. Stevenson on Fiction*, 57.
88 Stevenson, "A Gossip on Romance," 58.
89 Stevenson, "A Humble Remonstrance," 57–8.

90 Stevenson, "A Humble Remonstrance," 84, 87.
91 Michel Foucault, *The Birth of the Clinic: An Archaeology of Medical Perception*, trans. A. M. Sheridan Smith (New York: Vintage Books, 1994), 31.
92 Elaine Showalter, *The Female Malady: Women, Madness, and English Culture, 1830–1980* (New York: Pantheon, 1985), 3.
93 Hacking, "*Automatisme Ambulatoire*": 34.
94 Showalter, *The Female Malady*, 4.
95 Jan Goldstein, "The Uses of Male Hysteria: Medical and Literary Discourse in Nineteenth-Century France," *Representations* 34 (1991): 134–65, 138.
96 Ian Hacking, "Two Souls in One Body," *Critical Inquiry* 17.4 (Summer 1991): 838–67, 839.

## 2 BRAM STOKER'S *DRACULA* AND CEREBRAL AUTOMATISM

1 Bram Stoker, *Dracula*, ed. Nina Auerbach and David Skal (New York: W. W. Norton and Company, 1997), 94. Further references to this edition will be cited parenthetically within the text.
2 J. Hillis Miller, *The Disappearance of God: Five Nineteenth-Century Writers* (Champaign: University of Illinois Press, 2000), 1–16.
3 Anne Harrington, *Medicine, Mind, and the Double Brain: A Study in Nineteenth-Century Thought* (Princeton University Press, 1987), 10.
4 William James, *The Principles of Psychology*, 2 vols. (New York: Henry Holt, 1923), 1: 215n.
5 *British Cyclopedia* (1838), s.v. "Psychology," quoted in Rick Rylance, *Victorian Psychology and British Culture, 1850–1880* (New York: Oxford University Press, 2000), 23.
6 Rylance, *Victorian Psychology*, 21.
7 James, *Principles of Psychology*, 1: 14.
8 Rylance, *Victorian Psychology*, 71.
9 For a balanced overview of "the Automaton-Theory," see James, *Principles of Psychology*, 1: 128–44.
10 The concept of cerebral automatism or epiphenomenalism likely originated in medicine with Huxley, Clifford, *et al.* during the late-Victorian period. The concept was imported into philosophy later on. *Stanford Encyclopedia of Philosophy*, s.v. "epiphenomenalism."
11 Thomas Henry Huxley, "On the Hypothesis That Animals Are Automata, and Its History," in *Method and Results: Essays by Thomas Henry Huxley* (New York: D. Appleton and Co., 1901), 199–250, 216.
12 Huxley, "On the Hypothesis That Animals Are Automata," 211.
13 Huxley, "On the Hypothesis That Animals Are Automata," 220.
14 Huxley, "On the Hypothesis That Animals Are Automata," 230.
15 Huxley, "On the Hypothesis That Animals Are Automata," 226.
16 James, *Principles of Psychology*, 1: 129.
17 Huxley, "On the Hypothesis That Animals Are Automata," 245–6.

18 Robert M. Young, *Mind, Brain and Adaptation in the Nineteenth Century: Cerebral Localization and Its Biological Context from Gall to Ferrier* (New York: Oxford University Press, 1990), 233.
19 Joseph LeDoux, *Synaptic Self: How Our Brains Become Who We Are* (New York: Penguin, 2002), 18.
20 LeDoux, *Synaptic Self*, 14.
21 William Kingdon Clifford, "Body and Mind," in *Lectures and Essays by the late William Kingdon Clifford, F.R.S.*, ed. Leslie Stephen and Fredrick Pollock (London: Macmillan, 1879), II: 31–70, 56.
22 Clifford, "Body and Mind," 58.
23 Oliver Wendell Holmes, Sr., "Crime and Automatism," *Atlantic Monthly* 35 (1875): 466–81, 481.
24 Young, *Mind, Brain and Adaptation*, 211.
25 Jenny Bourne Taylor and Sally Shuttleworth, eds., *Embodied Selves: An Anthology of Psychological Texts, 1830–1890* (New York: Oxford University Press, 1998), 95.
26 See, for instance, William Benjamin Carpenter, "The Power of the Will over Mental Action," in Taylor and Shuttleworth, *Embodied Selves*, 95–101.
27 William Benjamin Carpenter, "On the Doctrine of Human Automatism," *Contemporary Review* 25 (1875): 397–416, 397.
28 Carpenter, "On the Doctrine of Human Automatism": 398.
29 William Benjamin Carpenter, "On the Doctrine of Human Automatism, Part II," *Contemporary Review* 25 (1875): 940–62, 942–3 (emphasis in original).
30 Rylance, *Victorian Psychology*, 75.
31 Patrick Brantlinger, "The Gothic Origins of Science Fiction," *NOVEL: A Forum on Fiction* 14.1 (1980): 30–43, 31.
32 William Hughes, *Bram Stoker:* Dracula, *A Reader's Guide to Essential Criticism* (New York: Palgrave, 2009), 66. Chapter 2 of Hughes's volume contains an overview of *Dracula* criticism relating to late-Victorian physiology and medicine (46–76).
33 On degeneration and criminology in *Dracula*, see Ernest Fontana, "Lombroso's Criminal Man and Stoker's *Dracula*," *Victorian Newsletter* 66 (1984): 25–7; and Daniel Pick, "'Terrors of the Night': *Dracula* and 'Degeneration' in the Late Nineteenth Century," *Critical Quarterly* 30 (1988): 71–87. On *Dracula*'s engagement with sexology, see Robert Mighall, *A Geography of Victorian Gothic Fiction: Mapping History's Nightmares* (New York: Oxford University Press, 1999), 210–47. On blood and menstruation in the novel, see David Flood, "Blood and Transfusion in Bram Stoker's *Dracula*," *University of Mississippi Studies in English*, New Series, 7 (1989): 180–92; and Marie Mulvey-Roberts, "*Dracula* and the Doctors: Bad Blood, Menstrual Taboo and the New Woman," in William Hughes and Andrew Smith, eds., *Bram Stoker: History, Psychoanalysis and the Gothic* (New York: Macmillan, 1998), 78–95.
34 See David Glover, *Vampires, Mummies and Liberals: Bram Stoker and the Politics of Popular Fiction* (Durham, NC: Duke University Press, 1996), 76–9;

William Hughes, "'So Unlike the Normal Lunatic': Abnormal Psychology in Bram Stoker's *Dracula*," *University of Mississippi Studies in English* 11/12 (1993–5): 1–10; William Hughes, *Beyond Dracula: Bram Stoker's Fiction and Its Cultural Context* (New York: Macmillan, 2000), 141–51; and John L. Greenway, "'Unconscious Cerebration' and the Happy Ending of *Dracula*," *Journal of Dracula Studies* 4 (2002): 1–9.

35 Carpenter held up publication of the 1874 edition of his *Principles of Mental Physiology* so that he could take account of Ferrier's cerebral localization research. In an appendix entitled "Dr. Ferrier's Experimental Researches on the Brain," Carpenter lauded Ferrier's methods and conclusions but misinterpreted them somewhat. Young, *Mind, Brain and Adaptation*, 109, 214–15.

36 Carpenter wrote, somewhat pathetically, "I have been publicly informed [that] I am getting old, and my brain is becoming ossified, so as no longer to be able to keep up with the advance of other brains." "On the Doctrine of Human Automatism, Part II," 944.

37 Carpenter, "On the Doctrine of Human Automatism," 416.

38 Hughes, *Bram Stoker*, 76.

39 Critics have made much of Stoker's fascination with these new technologies. See, for instance, Jennifer Wicke, "Vampiric Typewriting: Dracula and Its Media," *ELH* 59.2 (1992): 467–93; Jennifer Fleissner, "Dictation Anxiety: The Stenographer's Stake in *Dracula*," *Nineteenth-Century Contexts* 22 (2000): 417–55; and Laura Otis, *Networking: Communicating With Bodies and Machines in the Nineteenth Century* (Ann Arbor: University of Michigan Press, 2001), 194–219.

40 Barbara Belford, *Bram Stoker: A Biography of the Author of* Dracula (New York: Knopf, 1996), 23.

41 Paul Murray, *From the Shadow of* Dracula: *A Life of Bram Stoker* (London: Jonathan Cape, 2004), 18.

42 Murray, *From the Shadow of* Dracula, 27.

43 See Carol Senf, *Science and Social Science in Bram Stoker's Fiction* (Westport, CT: Greenwood Press, 2002).

44 Bram Stoker, *Personal Reminiscences of Henry Irving* (London: William Heinemann, 1906), 20; Murray, *From the Shadow of* Dracula, 33.

45 Belford, *Bram Stoker*, 34.

46 J. V. Luce, *Trinity College Dublin: The First 400 Years* (Trinity College Dublin Press, 1992), 103.

47 Luce, *Trinity College Dublin*, 103.

48 Senf, *Science and Social Science in Bram Stoker's Fiction*, 7.

49 Belford, *Bram Stoker*, 244.

50 Murray, *From the Shadow of* Dracula, 14.

51 Belford, *Bram Stoker*, 69.

52 "Musha, but Docther Wilde himself, Rest his sowl! couldn't have put it aisier to grip. It's a purfessionaler the young gintleman is intirely!" Bram Stoker, *The Snake's Pass* (Dingle, Ireland: Brandon, 1990), 56.

53 Robert Gray, *The Theory of Dreams: In Which an Inquiry Is Made into the Powers and Faculties of the Human Mind, as They Are Illustrated in the Most Remarkable Dreams*, 2 vols. (London: F. C. and J. Livingston, 1808), II: 30.
54 Gray, *The Theory of Dreams*, II: 41–2.
55 Gray, *The Theory of Dreams*, II: 43.
56 Gray, *The Theory of Dreams*, II: 62.
57 Sarah Lee, *Anecdotes on the Habits and Instincts of Animals* (New York: E. P. Dutton, 1852), 31.
58 Gray, *The Theory of Dreams*, II: 69.
59 Belford, *Bram Stoker*, 94.
60 Anne Stiles, Stanley Finger, and John Bulevich, "Somnambulism and Trance States in the Works of John William Polidori, Author of *The Vampyre*," *European Romantic Review* 21.6 (2010): 789–807.
61 "Obituary: Sir William Thornley Stoker, Bart., M.D.," *British Medical Journal* (June 15, 1912): 1399–400, 1399.
62 "Obituary: Sir William Thornley Stoker, Bart., M.D.": 1399.
63 John Fleetwood, *History of Medicine in Ireland* (Dublin: The Richview Press, 1951), 178.
64 See William Thornley Stoker *et al.*, "A Contribution to the Surgery of the Brain," *Dublin Journal of Medical Science* 90 (1890): 292–9, 487–95; and William Thornley Stoker, "On a Case of Subcranial Haemorrhage Treated by Secondary Trephining," *Annals of Surgery* 7 (1888): 401–9.
65 William Thornley Stoker, "On a Case of Subcranial Haemorrhage": 407–8.
66 "Royal Commission on Vivisection: Inspection in Ireland," *British Medical Journal* (March 2, 1907): 507–12, 507.
67 William Thornley Stoker, "Memorandum on Head Injuries," *c.*1890–6, EL3 f.S874d MS, Dracula: Notes and Outline, Rosenbach Museum and Library, Philadelphia, PA.
68 William Thornley Stoker, "Memorandum on Head Injuries."
69 William Thornley Stoker, "On a Case of Subcranial Haemorrhage": 403.
70 On the risks of late-Victorian psychosurgery, particularly the imperfect antiseptic techniques that resulted in postoperative infections, see Stanley Finger, *Minds Behind the Brain: A History of the Pioneers and Their Discoveries* (New York: Oxford University Press, 2000), 168, 172–4.
71 William Thornley Stoker, "Memorandum on Head Injuries."
72 Though Burdon-Sanderson is now best known for first measuring the electrical impulses that come from the heart, he also conducted experiments on animal brains that David Ferrier cited in his landmark work, *The Functions of the Brain* (New York: G. P. Putnam's Sons, 1876).
73 John L. Greenway, "Seward's Folly: *Dracula* as a Critique of 'Normal Science,'" *Stanford Literature Review* 3.2 (1986): 213–30.
74 Greenway, "Seward's Folly": 218.
75 Finger, *Minds Behind the Brain*, 167, 175.
76 Charles Sherrington, "Sir David Ferrier, 1843–1928," paraphrased in Young, *Mind, Brain and Adaptation*, 236.

77 Young, *Mind, Brain and Adaptation*, 234.
78 My overview of Ferrier's experiments and his 1881 trial is drawn from several sources, in addition to Ferrier's own works. These secondary sources include Finger, *Minds Behind the Brain*, 162–75; Young, *Mind, Brain and Adaptation*, 234–48; and Laura Otis, "Howled out of the Country: Wilkie Collins and H. G. Wells Retry David Ferrier," in Anne Stiles, ed., *Neurology and Literature, 1860–1920* (New York: Palgrave, 2007), 27–51. I am particularly indebted to Otis for her engaging description of the cultural impact of Ferrier's work.
79 Ferrier, *The Functions of the Brain*, 143.
80 Ferrier, *The Functions of the Brain*, 144.
81 David Ferrier, "Experimental Researches in Cerebral Physiology and Pathology" (1873), quoted in Otis, "Howled out of the Country," 30.
82 Otis, "Howled out of the Country," 31.
83 Finger, *Minds Behind the Brain*, 167.
84 David Ferrier, *The Localization of Cerebral Disease* (New York: G. P. Putnam's Sons, 1879), 21.
85 On Goltz's role at the International Medical Congress, see Finger, *Minds Behind the Brain*, 155–8.
86 Young, *Mind, Brain and Adaptation*, 240.
87 Finger, *Minds Behind the Brain*, 158.
88 Ferrier, *The Functions of the Brain*, 125.
89 Otis, "Howled out of the Country," 35.
90 Frances Power Cobbe, *The Modern Rack: Papers on Vivisection* (London: Swan Sonnenschein, 1889), 219.
91 Cobbe, *The Modern Rack*, 172.
92 Ferrier, *The Functions of the Brain*, 129–30.
93 Richard French, *Antivivisection and Medical Science in Victorian Society* (Princeton University Press, 1975), 202–3.
94 Coral Lansbury, *The Old Brown Dog: Women, Workers and Vivisection in Edwardian England* (Madison: University of Wisconsin Press, 1985), 153.
95 The latest examples of such activism include glossy advertising campaigns from PETA (People for the Ethical Treatment of Animals), and the "animal studies" trend that is currently sweeping humanities disciplines. Despite the increased visibility of animal rights concerns among scholars and the general public, it seems to me unlikely that renewed antivivisection efforts will make much headway, given the tremendous financial and cultural authority of science in the twenty-first century.
96 French, *Antivivisection and Medical Science*, 201.
97 Otis, "Howled out of the Country," 37–47.
98 Finger, *Minds Behind the Brain*, 161.
99 Gustav Fritsch and Eduard Hitzig, "On the Electrical Excitability of the Cerebrum" (1870), quoted in Young, *Mind, Brain and Adaptation*, 228.
100 Ferrier, *The Functions of the Brain*, 256 (emphasis in original).

101 Ferrier, *The Functions of the Brain*, 256.
102 Electrical induction is "the phenomenon that produces an electrical charge in an adjacent or proximate circuit, apparently acting across empty space." Roger Luckhurst, *The Invention of Telepathy, 1870–1901* (New York: Oxford University Press, 2002), 75.
103 Emily Gerard, *The Land Beyond the Forest* (1888), quoted in Stoker, *Dracula*, 212n.
104 Roslynn D. Haynes, *From Faust to Strangelove: Representations of the Scientist in Western Literature* (Baltimore: Johns Hopkins University Press, 1994), 3.
105 The cranial structure and facial appearance of geniuses will be more thoroughly explored in Chapter 4.
106 Stoker, *Dracula*, 171n.
107 Finger, *Minds Behind the Brain*, 182.
108 Finger, *Minds Behind the Brain*, 191.
109 Finger, *Minds Behind the Brain*, 191.
110 Anna Kingsford, quoted in Lansbury, *The Old Brown Dog*, 87.
111 Lansbury, *The Old Brown Dog*, 82.
112 Cobbe, *The Modern Rack*, 241.
113 Cobbe, *The Modern Rack*, 244.
114 Cobbe, *The Modern Rack*, 247. Cobbe's nationalist prejudices seem hardly fair – English physiologists were at least as callous toward their animal subjects as their French counterparts, as Burdon-Sanderson's *Handbook* amply demonstrates.
115 Stoker, *The Snake's Pass*, 59.
116 Greenway, "Seward's Folly": 222–3.
117 Stoker, *Dracula*, 171n.
118 Leonard Ashley, "Introduction" to Edmund Gurney *et al.*, *Phantasms of the Living* (Gainesville, FL: Scholars' Facsimiles and Reprints, 1970), v.
119 "Members and Associates," *Proceedings of the Society for Psychical Research* 19 (1907): 463.
120 Quoted in Renée Haynes, *The Society for Psychical Research, 1882–1982: A History* (London: Macdonald, 1982), 6.
121 Haynes, *The Society for Psychical Research*, 1.
122 Haynes, *The Society for Psychical Research*, 1.
123 Janet Oppenheim, *The Other World: Spiritualism and Psychical Research in England, 1850–1914* (Cambridge University Press, 1985), 152.
124 Oppenheim, *The Other World*, 154.
125 Stoker, *Personal Reminiscences of Henry Irving*, 396.
126 "How to Humanize the Landscape," *Punch* (March 15, 1911): 195.
127 Belford, *Bram Stoker*, 165.
128 Belford, *Bram Stoker*, 315.
129 Greenway, "Seward's Folly": 213–30.
130 Christopher Frayling, "Bram Stoker's Working Papers for *Dracula*," in Stoker, *Dracula*, 339–50, 343.

131 John Greenway makes a similar point, stating that "Seward resembles Ferrier more than Carpenter in his treatment of his patient [Renfield]." "'Unconscious cerebration' and the Happy Ending of *Dracula*": 3.
132 Huxley, "On the Hypothesis That Animals Are Automata," 231.
133 Frederic Myers, *Human Personality and Its Survival of Bodily Death*, ed. Leopold Hamilton Myers (London: Longmans, Green, and Co., 1918), 168.
134 Oppenheim, *The Other World*, 153.
135 Nina Auerbach and David Skal, "Preface" to Stoker, *Dracula*, ix–xiv, xiii.
136 LeDoux, *Synaptic Self*, 12.
137 Nina Auerbach, *Our Vampires, Ourselves* (University of Chicago Press, 1995), 1–10.
138 Mark McGurl, "Zombie Renaissance: Eating Your Brains," *N+1* (April 27, 2010): http://nplusonemag.com/the-zombie-renaissance-r-n (accessed October 25, 2010).
139 In "Zombie Renaissance," McGurl describes zombies and vampires as fundamentally different. In his view, vampires are the celebrities of the monster universe, embodying "dreams of exalted individuality," whereas zombies are the "monster lumpenproletariat" who embody the neuroscientific idea that we are the sum of our reflexes and synaptic connections. By contrast, I see zombies and vampires existing along the same continuum. I feel that modern vampires are still capable of expressing concerns about biological determinism, especially the idea that we are purely physical beings without souls or independent volition.
140 Stephenie Meyer, *New Moon* (New York: Little, Brown and Company, 2006), 69.

## 3 PHOTOGRAPHIC MEMORY IN THE WORKS OF GRANT ALLEN

1 Grant Allen, *Physiological Aesthetics* (New York: D. Appleton and Co., 1877), 19.
2 Grant Allen, "A Thinking Machine," *The Gentleman's Magazine* 260 (1886): 30–41, 30.
3 Allen, *Physiological Aesthetics*, 22, 15.
4 *Encyclopaedia Britannica*, 11th edn. (1910), s.v. "René Descartes." Some twentieth-century scholars have suggested that Victorian supporters of vivisection misunderstood Descartes's views about whether animals feel pain. See, for example, Peter Harrison, "Descartes on Animals," *The Philosophical Quarterly* 42.167 (1992): 219–27.
5 For an overview of Victorian theories of cerebral automatism, see William James, *The Principles of Psychology*, 2 vols. (New York: Henry Holt, 1923), 1: 128–44.
6 Alexander Bain, *Mental Science: A Compendium of Psychology, and the History of Philosophy* (New York: D. Appleton, 1886), 75.

7 Peter Morton, *"The Busiest Man in England": Grant Allen and the Writing Trade, 1875–1900* (New York: Palgrave, 2005), 111.
8 Only a handful of critics have analyzed Allen's female detective novels at length. Recent studies include Chris Willis, "The Detective's *Doppelgänger*: Conflicting States of Female Consciousness in Grant Allen's Detective Fiction," in William Greenslade and Terence Rodgers, eds., *Grant Allen: Literature and Cultural Politics at the* Fin de Siècle (Burlington, VT: Ashgate, 2005), 143–53; Phyllis Rozendal, "Grant Allen," in *British Mystery Writers, 1860–1919*, vol. LXX of Bernard Benstock and Thomas F. Staley, eds., *Dictionary of Literary Biography* (Detroit: Gale, 1988), 3–13; and Joseph Kestner, *Sherlock's Sisters: The British Female Detective, 1864–1913* (Burlington, VT: Ashgate, 2003), 119–34 and 157–69.
9 Ronald Thomas, *Detective Fiction and the Rise of Forensic Science* (Cambridge University Press, 1999), 3.
10 Lawrence Frank, *Victorian Detective Fiction and the Nature of Evidence: The Scientific Investigations of Poe, Dickens, and Doyle* (New York: Palgrave, 2003), 162.
11 *Miss Cayley's Adventures* and *Hilda Wade* were both published serially in *The Strand* between March 1898 and February 1900, while *What's Bred in the Bone* was serialized in George Newnes's *Tit-bits* during 1890–1. *Recalled to Life* is atypical of Allen's fiction in that it was never serialized. Morton, *"The Busiest Man in England"*, 136, 182; Willis, "The Detective's *Doppelgänger*," 145.
12 Elizabeth Carolyn Miller, *Framed: The New Woman Criminal in British Culture at the* Fin de Siècle (Ann Arbor: University of Michigan Press, 2008), 25–51.
13 Conan Doyle began a course of study in ophthalmology in Vienna in late 1890, after becoming interested in the field while working at the Portsmouth and South Hants Eye and Ear Infirmary. Thomas, *Detective Fiction*, 167.
14 Miller, *Framed*, 32.
15 Arthur Conan Doyle, *A Study in Scarlet*, in *A Study in Scarlet and Sign of the Four* (New York: Berkley, 1963), 23.
16 Arthur Conan Doyle, "A Scandal in Bohemia," in *The New Annotated Sherlock Holmes*, ed. Leslie Klinger (New York: W. W. Norton and Company, 2005), 1: 5.
17 Thomas, *Detective Fiction*, 171.
18 William Paley, from *Natural Theology* (1802), in *Darwin*, ed. Philip Appleman (New York: W. W. Norton and Company, 2001), 41–4, 44.
19 Charles Darwin, from *The Origin of Species*, in *Darwin*, ed. Appleman, 95–174, 144.
20 Darwin, *Origin*, 144–5.
21 Darwin, *Origin*, 145.
22 See, for instance, René Descartes's views on vision in *Treatise on Man* (1664), particularly several illustrations (figures 64, 66, and 67) that feature the eyes communicating directly with the pineal gland. Descartes famously saw the

pineal gland as the seat of the soul. René Descartes, *The World and Other Writings*, ed. Stephen Gaukroger (Cambridge University Press, 1998), 154, 157, 158.

23 Conan Doyle famously declared his belief in spiritualism in 1916, and later wrote *The History of Spiritualism* (1926). He also argued for the existence of fairies in such works as *The Coming of the Fairies* (1921).

24 Morton, *"The Busiest Man in England"*, 3.

25 Richard Proctor, "Letters Received and Short Answers," *Knowledge* 4 (November 30): 338; quoted in Bernard Lightman, *Victorian Popularizers of Science: Designing Nature for New Audiences* (University of Chicago Press, 2007), 316.

26 Morton, *"The Busiest Man in England"*, 96.

27 Grant Allen, *Recalled to Life* (Hard Press, 2005), 5, 7. Further references to this source will be cited parenthetically.

28 In a review of Ribot's later work, *L'Hérédité Psychologique* (1882), Allen writes that "it is no disparagement to his work to say that it is neither so novel in thought nor so generally important as his *Maladies de la Mémoire* … in which perhaps he did himself full justice for the first time as an original psychological thinker." Grant Allen, "Review of *L'Hérédité Psychologique*, by Th. Ribot," *Mind* 7.27 (July 1882): 413–15, 414.

29 As discussed in Chapter 1, nineteenth-century case studies of dual personality often mention the patient entering trance-like or hypnoid states. For more modern views on dissociation – including its role in multiple personality disorder (now called dissociative identity disorder) – see "What is Dissociation?" International Society for the Study of Trauma and Dissociation, 2010: www.isst-d.org/education/faq-dissociation.htm#dissocamn (accessed July 28, 2010).

30 Douwe Draaisma, *Metaphors of Memory: A History of Ideas About the Mind*, trans. Paul Vincent (New York: Cambridge University Press, 2000), 104, 3.

31 Laura Otis, "The Metaphoric Circuit: Organic and Technological Communication in the Nineteenth Century," *Journal of the History of Ideas* 63.1 (January 2002): 105–28.

32 "Alexander Graham Bell Biography," *Encyclopedia of World Biography* (2007): www.notablebiographies.com/Ba-Be/Bell-Alexander-Graham.html (accessed November 10, 2008).

33 Véronique Campion-Vincent, "The Tell-Tale Eye," *Folklore* 110 (1999): 13–24, 13. For a more detailed account of Böll's experiments and their scientific implications, see John E. Dowling, *Neurons and Networks: An Introduction to Behavioral Neuroscience* (Cambridge, MA: Harvard University Press, 2001), 194.

34 Andrea Goulet, "Retinal Fictions: Villiers, Leroux, and Optics at the Fin-de-Siècle," *Nineteenth-Century French Studies* 34.1–2 (2005–6): 107–20, 109.

35 Campion-Vincent, "The Tell-Tale Eye": 14.

36 See Grant Allen, *The Colour Sense: Its Origin and Development* (London: Truebner, 1879), 120, and "Mr. G. S. Hall on the Perception of Colour," *Mind* 4.14 (April 1879): 267–8, 268.

37 Campion-Vincent, "The Tell-Tale Eye": 19.
38 Campion-Vincent, "The Tell-Tale Eye": 15.
39 Campion-Vincent, "The Tell-Tale Eye": 14.
40 Campion-Vincent, "The Tell-Tale Eye": 20.
41 Roger Luckhurst has suggested that Allen's views on pseudoscience were more conflicted than the author himself realized: "Allen, for all his Spencerian credentials, fostered the kinds of cross-fertilizations between orthodox and marginal science that encouraged the confusions he despised." *The Invention of Telepathy* (New York: Oxford University Press, 2002), 199.
42 Théodule Ribot, *Diseases of Memory: An Essay in the Positive Psychology*, 5th edn. (London: Kegan Paul, 1906), 95.
43 On composition photography, see Daniel Novak, *Realism, Photography, and Nineteenth-Century Fiction* (Cambridge University Press, 2008), 1–35.
44 On the phonograph and telegraphic switchboard as alternative metaphors for memory, see Draaisma, *Metaphors of Memory*, 85–90, 121, 187.
45 The ups and downs of Allen's scientific career are recorded in Morton, *"The Busiest Man in England"*, 95–110. Though skeptical of Allen's attempts at original science, Morton concludes that "as a scientific popularizer, synthesizer, and middleman, Allen was without peer in his own day" (106).
46 Suzy Anger, "Review of *'The Busiest Man in England': Grant Allen and the Writing Trade, 1875–1900*, and: *Grant Allen: Literature and Cultural Politics at the Fin de Siècle*," *Victorian Studies* 49.2 (Winter 2007): 362–5, 364.
47 In the preface to *Strange Stories*, for instance, Allen refers to himself as "by trade a psychologist and scientific journeyman." Grant Allen, *Strange Stories* (London: Chatto & Windus, 1884), iii. In his preface to *The Colour Sense*, meanwhile, Allen describes himself as a "comparative psychologist" (vii).
48 Grant Allen, "Personal Reminiscences of Herbert Spencer," *The Forum* 35 (1903–4): 610–28, 614.
49 Herbert Spencer, *The Principles of Psychology*, 2 vols. (New York: D. Appleton, 1897), 1: 358.
50 Spencer, *Principles of Psychology*, 1: 363. These examples are cited in Grant Allen, "The Dog's Universe," *The Gentleman's Magazine* 249 (1880): 287–301. Allen writes: "Mr. Herbert Spencer has pointed out that intelligence varies amongst animals generally in a rough proportion to their special organs of touch and prehension" (287–8).
51 Charles Darwin, from *The Descent of Man* (1871), in *Darwin*, ed. Appleman, 175–254, 188.
52 Allen, *Physiological Aesthetics*, 140.
53 Grant Allen, "Sight and Smell in Vertebrates," *Mind* 6.24 (October 1881): 453–71, 469.
54 Allen, "Sight and Smell in Vertebrates": 453.
55 Allen, "The Dog's Universe": 294–5.
56 Allen, "Sight and Smell in Vertebrates": 470, 465.
57 Allen, "The Dog's Universe": 293.
58 Allen, "The Dog's Universe": 291.

59 Elsewhere, Allen expressed a more qualified approval of vivisection: "Vivisection for genuine scientific purposes in responsible hands, one can understand and tolerate, even though lacking the heart for it oneself." Grant Allen, "Slugs and Snails," in *The Evolutionist at Large* (London: Chatto & Windus, 1881), 48–58, 48.

60 Morton, *"The Busiest Man in England,"* 4.

61 Richard le Gallienne, "Grant Allen," in *Attitudes and Avowals, With Some Retrospective Reviews* (London: John Lane, 1910), 167–211, 173.

62 Le Gallienne, "Grant Allen," 172–3. On Allen's microscope, see also Morton, *"The Busiest Man in England,"* 81–2.

63 The existence and medical status of "photographic memory" is highly controversial. See Alan Searleman, "Is there such a thing as a photographic memory? And if so, can it be learned?" *Scientific American* (March 12, 2007): www.sciam.com/article.cfm?id=is-there-such-a-thing-as (accessed November 2, 2008).

64 Henry Maudsley, *The Physiology of Mind* (New York: D. Appleton, 1878), 517–18.

65 Ribot, *Diseases of Memory*, 133.

66 Ribot, *Diseases of Memory*, 137. A "lightning calculator" is a person who can solve complex mathematical problems very rapidly, without pen and paper. On Victorian scientific studies of such individuals, see Draaisma, *Metaphors of Memory*, 129–32.

67 Allen praises "Dr. Maudsley" as a "leader … in the science of mind" in *Physiological Aesthetics*, 3.

68 George Frederick Drinka, *The Birth of Neurosis: Myth, Malady, and the Victorians* (New York: Simon & Schuster, 1984), 54. See Chapter 4 of the present volume for an extended discussion of the links between genius and insanity.

69 Draaisma, *Metaphors of Memory*, 129–33.

70 On Una's illness as a case of dual consciousness, see Willis, "The Detective's *Doppelgänger*," 150–1.

71 Félida's case is described in Ribot, *Diseases of Memory*, 102–3. Félida's bizarre symptoms were detailed at greater length by Allen's friend Richard Proctor in a series of articles for *Cornhill Magazine*, to which Allen also contributed. It is therefore practically certain that Allen knew about Félida X. and other cases of dual and multiplex personality discussed in the British press.

72 Ribot writes, "the difference between the victim of aphasia and the deaf-mute would then be, that one had forgotten what he had learned, and that the other had not yet been taught." *Diseases of Memory*, 163.

73 Allen died before he could complete *Hilda Wade*. At Allen's request, Conan Doyle finished the last two chapters for him. Morton, *"The Busiest Man in England,"* 182.

74 Grant Allen, *Hilda Wade: A Woman With Tenacity of Purpose* (New York: G. P. Putnam, 1900), 107.

75 Ribot, *Diseases of Memory*, 61.

76 Allen, *Hilda Wade*, 74.

77 Dr. Cumberledge, Hilda's future fiancé, constructs a device consisting of sequential photographs pasted inside a small wheel, also known as a zoetrope. When rapidly spinning, the wheel "made a living picture like a cinematograph." Allen, *Hilda Wade*, 325.

78 Hilda's father, Dr. Yorke-Bannerman, had "an astounding memory. Recollected every symptom of every patient he ever attended. And *such an eye!* ... Diagnosis? It was clairvoyance! A gift, no less. Knew what was the matter with you the moment he looked at you." Allen, *Hilda Wade*, 175.

79 Allen, "Sight and Smell in Vertebrates": 470.

80 Allen, "Sight and Smell in Vertebrates": 470.

81 Allen, "The Dog's Universe": 291.

82 H. G. Wells, *The War of the Worlds*, ed. Patrick Parrinder (New York: Penguin, 2005), 125.

83 Wells's intellectual debt to Allen has been the focus of scholarly debate. See, for example, David Y. Hughes, "A Queer Notion of Grant Allen's," *Science Fiction Studies* 25.2 (June 1998): 271–84.

84 Spencer, *The Principles of Psychology*, 1: 436.

85 Henry Maudsley, *The Physiology of Mind* (New York: D. Appleton, 1889), 213.

86 On cases of mistaken identity based on photographic "evidence," see Novak, *Realism, Photography, and Nineteenth-Century Fiction*, 48–50.

87 On the opposition between "'high' Victorian realism" and "the subversive 'other Victorian' literary underworlds of gothic, detective, and sensation fiction," see Thomas, *Detective Fiction*, 13. On overlaps between detective fiction and the Gothic, see Nils Clausson, "Degeneration, *Fin-de-Siècle* Gothic, and the Science of Detection: Arthur Conan Doyle's *The Hound of the Baskervilles* and the Emergence of the Modern Detective Story," *Journal of Narrative Theory* 35.1 (Winter 2005): 60–87.

88 Willis, "The Detective's *Doppelgänger*," 149.

89 Nicholas Dames, *Amnesiac Selves: Nostalgia, Forgetting, and British Fiction, 1810–1870* (New York: Oxford University Press, 2001), 169.

90 While Mr. Candy in *The Moonstone* experiences thorough and lasting memory loss, other characters, such as Laura Fairlie in *The Woman in White* (1860), experience partial memory lapses as a result of specific traumatic events (her placement in a mental asylum). Dames, *Amnesiac Selves*, 168–9.

91 Thomas, *Detective Fiction*, 7.

92 Richard Marsh, best known for his Gothic tale *The Beetle* (1897), also authored the female detective serial *Judith Lee: Some Pages from Her Life* (1912). Mary Elizabeth Braddon's *Thou Art the Man* (1894) also features at least one female detective. See Kestner, *Sherlock's Sisters*, 55–70 and 198–211.

93 Miller, *Framed*, 3.

94 Miller, *Framed*, 5.

95 In *Physiological Aesthetics*, Allen admits that he has "borrowed largely" from Helmholtz's *Handbuch der Physiologischen Optik*, in which Helmholtz discusses "mouches volantes" (floaters) among other curious visual phenomena

(x). Allen mentions Helmholtz's work on color vision in *The Colour Sense* (31, 223), where he also discusses "Kühne of Heidelberg" and his experiments on frog retinas (120).

96 On retinal after-images, which Allen terms "negative accidental images," see *Physiological Aesthetics*, 162–3. See also Allen, "Mr. G. S. Hall on the Perception of Colour": 267–8.

97 Goulet, "Retinal Fictions": 108.

98 Charlotte Perkins Gilman, "The Yellow Wallpaper," quoted in Goulet, "Retinal Fictions": 116.

99 Goulet, "Retinal Fictions": 116.

100 Clausson, "Degeneration, *Fin-de-Siècle* Gothic, and the Science of Detection": 64.

101 Valdine Clemens, *The Return of the Repressed*, quoted in Clausson, "Degeneration, *Fin-de-Siècle* Gothic, and the Science of Detection": 64.

102 Clausson, "Degeneration, *Fin-de-Siècle* Gothic, and the Science of Detection": 78.

103 On collage photography, retouching, and other "artistic" practices in Victorian photography, see Novak, *Realism, Photography, and Nineteenth-Century Fiction*, 1–62. On the similarities between Victorian photographic illusions and digital alterations of photographs, see 145–51.

104 Novak, *Realism, Photography, and Nineteenth-Century Fiction*, 4.

105 André Bazin, "The Ontology of the Photographic Image," in Alan Trachtenberg, ed., *Classic Essays on Photography* (New Haven, CT: Leete's Island, 1980), 237–44, 241.

106 Miller, *Framed*, 42.

107 Phillip Prodger, *Muybridge and the Instantaneous Photography Movement* (New York: Oxford University Press in association with the Iris and B. Gerald Cantor Center for Visual Arts at Stanford University, 2003), 43.

108 See Arthur Conan Doyle, *The Case for Spirit Photography* (1922).

109 In fact, many so-called "instantaneous" photographs were doctored, and not all had rapid exposure times. See Prodger, *Muybridge and the Instantaneous Photography Movement*, 38–43.

110 Brian Coe, "Eadweard Muybridge," in Stephen Herbert and Luke McKernan, eds., *Who's Who of Victorian Cinema* (London: British Film Institute, 1996), 99–100, 99.

111 Prodger, *Muybridge and the Instantaneous Photography Movement*, 25.

112 Walter Benjamin, "A Short History of Photography" (1931), in Trachtenberg, *Classic Essays on Photography*, 199–216, 203.

113 On Galton's interest in chronophotography and the potential medical applications of this technology see Prodger, *Muybridge and the Instantaneous Photography Movement*, 162–3 and 203.

114 This description of the zoopraxiscope is condensed from Prodger, *Muybridge and the Instantaneous Photography Movement*, 154–61. Prodger notes that zoopraxiscope discs were not usually made from actual photographs, which appeared unnaturally short and squat when projected using Muybridge's

device. Instead, most discs were made from "drawn and elongated copies of originals" (161).

115 Gordon Hendricks, *Eadweard Muybridge: The Father of the Motion Picture* (New York: Viking, 1975), vii.

116 Prodger, *Muybridge and the Instantaneous Photography Movement*, 223.

117 Kate Flint, "Throwing Light on the Moment: Flash Photography and Arresting Narratives" (paper presented at the 6th annual North American Victorian Studies Conference, New Haven, CT, November 14–16, 2008).

118 Allen, *Physiological Aesthetics*, 200. I am grateful to Kate Flint for drawing my attention to this passage.

119 Comparisons between flash photography and lightning were not uncommon in Victorian writing. For instance, in his well-known essay "The Stereoscope and the Stereograph" (1859), Oliver Wendell Holmes writes, "The time is perhaps at hand when a flash of light, as sudden and brief as that of the lightning … shall preserve the very instant of the shock of contact of the mighty armies that are even now gathering." Oliver Wendell Holmes, "The Stereoscope and the Stereograph" (1859) in Trachtenberg, *Classic Essays on Photography*, 82.

120 Ribot, *Diseases of Memory*, 107.

121 Ribot, *Diseases of Memory*, 49.

122 Jill Matus, "Emergent Theories of Victorian Mind Shock: From War and Railway Accident to Nerves, Electricity and Emotion," in Anne Stiles, ed., *Neurology and Literature, 1860–1920* (New York: Palgrave, 2007), 163–83, 172.

123 Spencer, *The Principles of Psychology*, I: 158.

124 Matus, "Victorian Mind Shock," 171.

125 The idea that nervous signals were electrical in nature was widely circulated in the late eighteenth century, but at this stage, electricity was still thought to be a fluid. On the discovery of electricity as the agent of nerve action, see Sidney Ochs, *A History of Nerve Functions: From Animal Spirits to Molecular Mechanisms* (New York: Cambridge University Press, 2004), 108–29.

126 Jill Matus, "Trauma, Memory, and Railway Disaster: The Dickensian Connection," *Victorian Studies* 43.3 (2001): 413–36, 413.

127 There are numerous reasons for the mid-Victorian proliferation of "railway spine" and "railway brain." Such physical diagnoses helped patients avoid the stigma of mental disorders like hysteria. These diagnoses also tended to help victims receive monetary compensation from railway corporations. See Matus, "Victorian Mind Shock," 170.

128 Matus, "Trauma, Memory, and Railway Disaster": 416.

129 Ribot, *Diseases of Memory*, 175–6. In 1877–8, Muybridge took a series of panoramic photos of San Francisco. See Paul Falconer, "Muybridge's Window to the Past: A Wet-plate View of San Francisco in 1877," *California History* 57 (1978): 130–57.

130 Ribot, *Diseases of Memory*, 85.

131 Ribot, *Diseases of Memory*, 47.

132 Hendricks, *Eadweard Muybridge*, 65–77.

133 In his essay "Muybridge in Motion: Travels in Art, Emotion, and Neurology," Arthur Shimamura attributes Muybridge's bizarre behavior to an injury to the anterior part of the frontal lobe of the brain, which helps to regulate emotions. Prodger, *Muybridge and the Instantaneous Photography Movement*, 123.

134 Prodger, *Muybridge and the Instantaneous Photography Movement*, 165–7.

135 Frank, *Victorian Detective Fiction*, 157.

136 Prodger, *Muybridge and the Instantaneous Photography Movement*, 161.

137 Josef Breuer and Sigmund Freud, *Studies on Hysteria*, trans. James Strachey (New York: Basic Books, 1895), 9, 17.

138 Allen cites this passage from Spencer's *The Principles of Psychology* in *Physiological Aesthetics*, 52.

139 Herbert Spencer, *The Principles of Psychology*, 2 vols. (New York: D. Appleton and Co, 1906), II: 431n.

140 Ribot, *Diseases of Memory*, 120.

141 Ribot, *Diseases of Memory*, 66.

142 Ribot, *Diseases of Memory*, 12. It should be noted, however, that Ribot sometimes utilized the sorts of mechanical metaphors he deplored, as in the above example where shallow minds are described as "mere machines."

143 The phrase "Determinist army" comes from William Benjamin Carpenter, "On the Doctrine of Human Automatism," *Contemporary Review* 25 (1875): 397–416, 397.

## 4 H. G. WELLS AND THE EVOLUTION OF THE MAD SCIENTIST

1 H. G. Wells, "The Man of the Year Million," paraphrased in *Punch*, "1,000,000 A.D.," November 25, 1893, 250.

2 H. G. Wells, "The Man of the Year Million," in *H. G. Wells: Journalism and Prophecy 1893–1946*, ed. W. Warren Wagar (Boston: Houghton Mifflin, 1964), 1–9, 9.

3 Roslynn D. Haynes, "Wells's Debt to Huxley and the Myth of Dr. Moreau," *Cahiers Victoriens et Edouardiens* 13 (1981): 31–41.

4 Jean Baptiste Lamarck, *Zoological Philosophy: An Exposition With Regard to the Natural History of Animals* (University of Chicago Press, 1984), 113.

5 Daniel Pick, *Faces of Degeneration: A European Disorder, c.1848–1918* (Cambridge University Press, 1989), 100–1.

6 George Becker, *The Mad Genius Controversy: A Study in the Sociology of Deviance* (Beverly Hills: Sage Publications, 1978), 85.

7 Anne Harrington, *Medicine, Mind, and the Double Brain: A Study in Nineteenth-Century Thought* (Princeton University Press, 1987), 35.

8 Harrington, *Medicine, Mind, and the Double Brain*, 35.

9 Harrington, *Medicine, Mind, and the Double Brain*, 42–3.

10 Paul Broca (1861), quoted in Harrington, *Medicine, Mind, and the Double Brain*, 42.
11 Stephen J. Gould, *The Mismeasure of Man* (New York: W. W. Norton and Company, 1981), 73–112.
12 Gould, *The Mismeasure of Man*, 149.
13 Gould, *The Mismeasure of Man*, 74.
14 H. G. Wells, Julian S. Huxley, and G. P. Wells, *The Science of Life* (New York: Doubleday, 1934), 794.
15 Wells referred to his early works as "scientific romances," but he was not the first to use the term, which originated with the lesser-known writer Charles Howard Hinton. Margaret Atwood, "Introduction" to H. G. Wells, *The Island of Dr. Moreau*, ed. Patrick Parrinder (New York: Penguin, 2005), xiii–xxvii, xvii.
16 H. G. Wells, "Introduction" to *The Scientific Romances of H. G. Wells* (London: Victor Gollancz, 1933), viii.
17 Wells, "Introduction" to *The Scientific Romances of H. G. Wells*, viii (emphasis in original).
18 Laura Otis, "Howled out of the Country: Wilkie Collins and H. G. Wells Retry David Ferrier," in Anne Stiles, ed., *Neurology and Literature, 1860–1920* (New York: Palgrave, 2007), 42.
19 There has recently been some revival of interest in Lamarckism as it relates to the emergent field of epigenetics. See Kara Rogers, "The Rebirth of Lamarckism (the Rise of Epigenetics)," *Encyclopaedia Britannica Blog* (February 25, 2009): www.britannica.com/blogs/2009/02/the-rebirth-of-lamarckism-the-rise-of-epigenetics (accessed November 22, 2010). See also Sharon Begley, "The Sins of the Fathers, Take 2," *Newsweek* (January 17, 2009): www.newsweek.com/2009/01/16/the-sins-of-the-fathers-take-2.html (accessed November 22, 2010).
20 H. G. Wells, "Review of *The Human Machine*, by John Ferguson Nisbet," *The Academy* (May 6, 1899): 502.
21 John Ferguson Nisbet, *The Insanity of Genius and the General Inequality of Human Faculty Physiologically Considered*, 6th edn. (New York: Charles Scribner, 1912), 57.
22 "Review of *The Insanity of Genius and the General Inequality of Human Faculty, Physiologically Considered*, by J. F. Nisbet," *Mind* 16 (1891): 541.
23 Nisbet, *The Insanity of Genius*, 325.
24 Nisbet, *The Insanity of Genius*, xv.
25 Nisbet, *The Insanity of Genius*, 57. See also chapters I and II, where Nisbet discusses cerebral localization in detail.
26 Nisbet, *The Insanity of Genius*, 85.
27 For a comprehensive bibliography of literature on genius and insanity, see Wilhelm Lange-Eichbaum, *Genie, Irrsinn und Ruhm*, 7th edn., 12 vols. (Munich: Ernst Reinhardt Verlag, 1986), I: 237–53.
28 Nancy Andreasen, *The Creative Brain: The Science of Genius* (New York: Penguin, 2006), 84–5.

29 Seneca attributes this remark to Aristotle in *"De Tranquillitate Animi"* (On Tranquility of Mind), in *Moral Essays* XVII: 10.
30 This etymology was compiled from two sources: "Genius" in the *Oxford English Dictionary*, 2nd edn. (1989); and Andreasen, *The Creative Brain*, 6–7.
31 Frederick Burwick, *Poetic Madness and the Romantic Imagination* (University Park: Pennsylvania State University Press, 1996), 12.
32 Michel Foucault, *The Birth of the Clinic: An Archaeology of Medical Perception*, trans. A. M. Sheridan Smith (New York: Vintage Books, 1994), 34.
33 *Encyclopaedia Britannica Online*, s.v. "Quételet, Adolphe": www.search.eb.com/eb/article-9062246 (accessed November 22, 2010).
34 Andreasen, *The Creative Brain*, 82.
35 Daniel Nettle, *Strong Imagination: Madness, Creativity and Human Nature* (New York: Oxford University Press, 2001), 37.
36 Nisbet, *The Insanity of Genius*, xxiii–xxiv.
37 Cesare Lombroso, *The Man of Genius* (New York: Garland Publishing, 1984), 73.
38 Roslynn D. Haynes, *From Faust to Strangelove: Representations of the Scientist in Western Literature* (Baltimore: Johns Hopkins University Press, 1994), 3.
39 Robert Philmus, "Introducing Moreau," in H. G. Wells, *The Island of Dr. Moreau: A Variorum Text*, ed. Robert Philmus (Athens: University of Georgia Press, 1993), xi–xlviii, xx–xxi.
40 James Sully, "Genius and Insanity," *The Nineteenth Century* 27 (1885): 948–69, 957.
41 Jan Goldstein, *Console and Classify: The French Psychiatric Profession in the Nineteenth Century* (New York: Cambridge University Press, 1987), 155–6, 159.
42 Becker, *The Mad Genius Controversy*, 29.
43 "Contrairement à ce qu l'on observe chez les hommes d'intelligence moyenne, le travail chez les hommes supérieurs, est tout spontané, et en quelque sorte, aussi peu volontaire que possible. Il est le résultat de l'entraînement, d'un besoin instinctif, d'une sorte d'appétit de l'intellect, que se fait sentir, on ne sait pourquoi … on dit avec raison que personne n'était moins libre dans ses travaux, de choisir son temps, son heure de travailler que les hommes dont nous parlons." Jacques Moreau, *La psychologie morbide dans ses rapports avec la philosophie de l'histoire ou de l'influence des névropathies sur le dynamisme intellectuel* (Paris: Librairie Victor Masson, 1859), 494–5.
44 George Frederick Drinka, *The Birth of Neurosis: Myth, Malady, and the Victorians* (New York: Simon & Schuster, 1984), 54.
45 Lombroso, *The Man of Genius*, viii.
46 Thomas Tyler, "Review of *The Man of Genius*, by Cesare Lombroso," *The Academy* (March 20, 1892): 303.
47 James Cowles Prichard, "Forms of Insanity," in Jenny Bourne Taylor and Sally Shuttleworth, eds., *Embodied Selves: An Anthology of Psychological Texts, 1830–1890* (New York: Oxford University Press, 1998), 252.
48 Lombroso, *The Man of Genius*, 14–15.

49 Robert Louis Stevenson, *Strange Case of Dr. Jekyll and Mr. Hyde*, ed. Katherine Linehan (New York: W. W. Norton and Company, 2003), 51.
50 Max Nordau, *Degeneration*, trans. George L. Mosse (Lincoln: University of Nebraska Press, 1993), 25.
51 Nordau, *Degeneration*, 541.
52 On this "ludicrous incident in the history of mental testing," see Gould, *The Mismeasure of Man*, 77, 183–8.
53 Francis Galton, *Hereditary Genius: An Inquiry into Its Laws and Consequences* (London: Watts and Co., 1950), x.
54 Francis Galton, *Hereditary Genius: An Inquiry into Its Laws and Consequences* (London: Macmillan and Co., 1869), 328. All further citations of *Hereditary Genius* refer to this edition.
55 Galton, *Hereditary Genius*, 328.
56 Pick, *Faces of Degeneration*, 100–1.
57 Peter Bowler, *The Eclipse of Darwinism: Anti-Darwinian Evolution Theories in the Decades around 1900* (Baltimore: Johns Hopkins University Press, 1983), 70–5.
58 Lombroso, *The Man of Genius*, v.
59 Sully, "Genius and Insanity": 966; Lombroso, *The Man of Genius*, 6.
60 Alexander Bain, "Education as a Science," *Mind* 2.5 (1877): 1–21, 7, 5.
61 Stanley Finger, *Minds Behind the Brain: A History of the Pioneers and Their Discoveries* (New York: Oxford University Press, 2000), 153.
62 Lombroso, *The Man of Genius*, vi.
63 See Philmus, "Introducing *Moreau*," xviii, xl; and John Glendening, "'Green Confusion': Evolution and Entanglement in H. G. Wells's *Island of Doctor Moreau*," *Victorian Literature and Culture* 30.2 (2002): 571–97, 579.
64 *Encyclopaedia Britannica Online*, s.v. "Weismann, August": www.search.eb.com/eb/article-9076462 (accessed November 22, 2010).
65 H. G. Wells, "Incidental Thoughts on a Bald Head," *Pall Mall Gazette* (March 1895), reprinted in Wells, *Certain Personal Matters* (London: T. Fisher Unwin, 1901), 106.
66 H. G. Wells, "Human Evolution, an Artificial Process," in *H. G. Wells: Early Writings in Science and Science Fiction*, ed. Robert Philmus (Berkeley: University of California Press, 1975), 211–19, 211.
67 Glendening, "'Green Confusion'"; 579.
68 "Review of *Difference in the Nervous Organisation of Man and Woman*, by Harry Campbell," *Mind* 16.63 (October 1891): 542.
69 Gustave Le Bon (1879), quoted in Gould, *The Mismeasure of Man*, 105.
70 Henry Maudsley, "Sex in Mind and Education," *Fortnightly Review* 15 (1874): 469–71.
71 Tyler, "Review of *The Man of Genius*," 304.
72 Jewish mad scientists in Victorian fiction include the eponymous main character of Julia Frankau's *Dr. Phillips: A Maida Vale Idyll* (1887), the villainous hypnotist Svengali in George du Maurier's *Trilby* (1894), and, possibly, Dr. Benjulia of Wilkie Collins's *Heart and Science* (1883). Although Benjulia's

ethnicity is never identified, his last name begins with the Hebrew prefix "Ben," meaning "son of."

73 H. G. Wells to Elizabeth Healey, February 7, 1901, in *The Correspondence of H. G. Wells*, ed. David Smith (London: Pickering & Chatto, 1998), I: 379. H. G. Wells to Francis Galton, Beatrice Webb, *et al.*, May 15, 1907 and June 15, 1907, *The Correspondence of H. G. Wells*, II: 149–52.

74 Wells, "Review of *The Human Machine*," 503.

75 A few previous critics compare the fictional and historical Moreaus, but none mention Wells's familiarity with *Morbid Psychology*. See Philmus, "Introducing Moreau," xviii, xli–xliii; Elaine Showalter, *Sexual Anarchy: Gender and Culture at the* Fin de Siècle (New York: Penguin, 1990), 178–9; Kelly Hurley, *The Gothic Body: Sexuality, Materialism, and Degeneration at the* Fin de Siècle (Cambridge University Press, 1996), 109.

76 Steven McLean, *The Early Fiction of H. G. Wells: Fantasies of Science* (New York: Palgrave, 2009), 3.

77 On eugenics in *The Time Machine* and *A Modern Utopia*, see McLean, *The Early Fiction of H. G. Wells*, 27–9, 151–88.

78 Galton, *Hereditary Genius*, 329.

79 Galton, *Hereditary Genius*, 331.

80 Galton, *Hereditary Genius*, 332.

81 Francis Galton, *English Men of Science: Their Nature and Nurture* (London: Macmillan and Co, 1874), 260.

82 H. G. Wells, *Anticipations of the Reaction of Mechanical and Scientific Progress upon Human Life and Thought* (New York: Harper & Brothers, 1902), 153.

83 Galton made recommendations about scientific education in *English Men of Science*, 235–60. Wells suggested improvements in scientific education in "Science, in School and after School," *Nature* 50 (September 1894): 525–6, and "Science Teaching: An Ideal and Some Realities," *Educational Times* 48 (January 1895): 23–9.

84 Wells, *Anticipations*, 290.

85 Wells, "The Man of the Year Million," 9.

86 The various titles following Dr. Nebogipfel's name suggest not only his scientific expertise, but also the possibility that he was modeled on Dr. Henry Jekyll, "M.D., D.C.L., LL.D., F.R.S., &c." Stevenson, *Dr. Jekyll and Mr. Hyde*, 13.

87 Bernard Bergonzi, *The Early H. G. Wells: A Study of the Scientific Romances* (Manchester University Press, 1961), 15, 114.

88 H. G. Wells, "Appendix I: *The Chronic Argonauts*," in *The Time Machine: An Invention. A Critical Text of the 1895 London First Edition, With an Introduction and Appendices*, ed. Leon Stover (Jefferson, NC: McFarland, 1996), 174–95, 178 (emphasis in original).

89 Wells, "Appendix I: *The Chronic Argonauts*," 192.

90 Wells, "Appendix I: *The Chronic Argonauts*," 192.

91 Wells, "Appendix I: *The Chronic Argonauts*," 191.

92 On Wells's unhappy experiences as a draper's assistant, see chapters 3 and 4 of his *Experiment in Autobiography: Discoveries and Conclusions of a Very Ordinary Brain (Since 1866)*, 2 vols. (Boston: Faber and Faber, 1984). These early ordeals also provided material for Wells's fiction, particularly *The History of Mr. Polly* (1910).
93 Wells, *Experiment in Autobiography*, 1: 38.
94 For an alternate reading, see Bergonzi, who argues that "Nebogipfel, Moreau, and Griffin are all variants of a single type, the scientist-as-alchemist, who is not disinterestedly concerned with knowledge for its own sake, but pursues it as a means of obtaining power." *The Early H. G. Wells*, 114.
95 Wells, "Appendix I: *The Chronic Argonauts*," 189.
96 Wells, *The Island of Doctor Moreau*, 17, 29.
97 Wells, *The Island of Doctor Moreau*, 79.
98 Wells, *Experiment in Autobiography*, 1: 309.
99 H. G. Wells, *The Island of Doctor Moreau: A Variorum Text*, 21–2.
100 On Moreau's Frankensteinian qualities, see also Chris Baldick, *In Frankenstein's Shadow: Myth, Monstrosity, and Nineteenth-Century Writing* (Oxford: Clarendon Press, 1987), 153–6.
101 Wells, *The Island of Doctor Moreau*, 48.
102 Lombroso, *The Man of Genius*, 57.
103 H. G. Wells, "Appendix I: The First Moreau," in *The Island of Doctor Moreau*, 136.
104 Collins's Benjulia comes to mind in this regard, as does Doctor Daniel Maclure of Sarah Grand's *The Beth Book* (1897).
105 Wells, *The Island of Doctor Moreau*, 34.
106 Otis, "Howled out of the Country," 42.
107 Otis, "Howled out of the Country," 44.
108 Wells, *Anticipations*, 168.
109 Wells, *Anticipations*, 167.
110 H. G. Wells, *The Works of H. G. Wells, Atlantic Edition, II*, quoted in McLean, *The Early Fiction of H. G. Wells*, 41.
111 For further parallels between Moreau and Wilde, see McLean, *The Early Fiction of H. G. Wells*, 41–2.
112 H. G. Wells, *The Invisible Man*, ed. Patrick Parrinder (New York: Penguin, 2005), 103.
113 Wells, *The Invisible Man*, 125.
114 Bergonzi likewise suggests that "[b]y the end of the book [Griffin] appears to be insane." *The Early H. G. Wells*, 118.
115 Wells, *The Invisible Man*, 20.
116 Robert Bogdan, *Freak Show: Presenting Human Oddities for Amusement and Profit* (University of Chicago Press, 1988).
117 Wells, *The Invisible Man*, 114.
118 Wells, *The Invisible Man*, 112.
119 Wells, *The Invisible Man*, 148.
120 Wells, *The Invisible Man*, 12, 8.

121 Wells, *The Invisible Man*, 22.
122 Wells, *The Invisible Man*, 15.
123 Sully, "Genius and Insanity": 963.
124 Bergonzi, *The Early H. G. Wells*, 125.
125 For a brilliant extended comparison of *The First Men in the Moon* and *Anticipations*, see McLean, *The Early Fiction of H. G. Wells*, 117–50.
126 H. G. Wells, *The War of the Worlds*, ed. Patrick Parrinder (New York: Penguin, 2005), 7.
127 Wells, *The War of the Worlds*, 9.
128 Bergonzi, *The Early H. G. Wells*, 134.
129 Wells, *The War of the Worlds*, 21.
130 Peter Kemp, *H. G. Wells and the Culminating Ape* (New York: St. Martin's, 1982), 24.
131 Wells, *The War of the Worlds*, 127.
132 Wells, *The War of the Worlds*, 125.
133 On Wells's fascination with tentacles and octopuses, see Kemp, *H. G. Wells and the Culminating Ape*, 25–9.
134 Wells, *The War of the Worlds*, 127.
135 Wells, *The War of the Worlds*, 126.
136 Gordon S. Haight, "H. G. Wells's 'Man of the Year Million,'" *Nineteenth-Century Fiction* 12.4 (March 1958): 323–6, 325.
137 Wells, *The War of the Worlds*, 125.
138 Wells, *The War of the Worlds*, 127.
139 Wells, "The Man of the Year Million," 8.
140 E. Ray Lankester, *Degeneration: A Chapter in Darwinism* (London: Macmillan and Co, 1880), 33.
141 Wells, *The War of the Worlds*, 46.
142 Wells, *The War of the Worlds*, 22.
143 Colin Manlove, "Charles Kingsley, H. G. Wells, and the Machine in Victorian Fiction," *Nineteenth-Century Literature* 48.2 (September 1993): 212–39, 239.
144 Manlove, "Charles Kingsley, H. G. Wells, and the Machine in Victorian Fiction": 235.
145 Wells, *The War of the Worlds*, 127.
146 W. Warren Wagar, *H. G. Wells: Traversing Time* (Middletown, CT: Wesleyan University Press, 2004), 57.
147 Wagar, *H. G. Wells*, 57.
148 Wells, *The War of the Worlds*, 67.
149 Patrick Parrinder, "How Far Can We Trust the Narrator of *The War of the Worlds*?" *Foundation* 77 (1999): 15–24.
150 Wells, "The Man of the Year Million," 9. See also *The Time Machine*, in which one group of human descendants, the Morlocks, live in a vast network of ventilated underground caves.
151 See Bergonzi, *The Early H. G. Wells*, 156–64; McLean, *The Early Fiction of H. G. Wells*, 117–88; and Wagar, *H. G. Wells*, 58.

152 Wagar, *H. G. Wells*, 60.
153 McLean, *The Early Fiction of H. G. Wells*, 121–4; Bergonzi, *The Early H. G. Wells*, 161–4.
154 McLean, *The Early Fiction of H. G. Wells*, 7.
155 H. G. Wells, *The First Men in the Moon*, ed. Patrick Parrinder (New York: Penguin, 2005), 184.
156 McLean, *The Early Fiction of H. G. Wells*, 150.
157 Wells, *The First Men in the Moon*, 182.
158 On insects in Wells's fiction, see Charlotte Sleigh, "Empire of the Ants: H. G. Wells and Tropical Entomology," *Science as Culture* 10 (2001): 33–71.
159 Wells, *The First Men in the Moon*, 186.
160 Wells, *The First Men in the Moon*, 182.
161 In *The Science of Life*, Wells and his co-authors wrote, "A check on the advance of *Homo Sapiens* is not necessarily the end of progressive evolution" (794–5).
162 Wells, *The First Men in the Moon*, 181.
163 Wells, *The First Men in the Moon*, 181.
164 Gould, *The Mismeasure of Man*, 97–8.
165 Wells, *The First Men in the Moon*, 181.
166 Wells, "The Man of the Year Million," 8.
167 Wells, *The First Men in the Moon*, 183.
168 Wells, *The First Men in the Moon*, 180.
169 Wells, *The First Men in the Moon*, 183.
170 Sleigh, "The Empire of the Ants," 58.
171 Wells, *The First Men in the Moon*, 192.
172 Wells, *The First Men in the Moon*, 191–2.
173 H. G. Wells, "Scepticism of the Instrument," *Mind* 13.51 (July 1904): 379–93, 391.
174 Roslynn D. Haynes, *H. G. Wells: Discoverer of the Future: The Influence of Science on His Thought* (London: Macmillan Press, 1980), 110–11.
175 Wells, *Experiment in Autobiography*, I: 20.
176 Wells, *Experiment in Autobiography*, I: 12.
177 See also Burwick, *Poetic Madness*; and Corinne Saunders and Jane Macnaughton, eds., *Madness and Creativity in Literature and Culture* (New York: Palgrave, 2005).
178 See Nancy Andreason, "Creativity and Mental Illness: Prevalence Rates in Writers and their First-Degree Relatives," *American Journal of Psychiatry* 144 (1987): 1288–92; Kay Redfield Jamison, "Mood Disorders and Patterns of Creativity in British Writers and Artists," *Psychiatry* 52 (1989): 125–34; and Joseph Schildkraut *et al.*, "Mind and Mood in Modern Art, II: Depressive Disorders, Spirituality, and Early Deaths in the Abstract Expressionist Artists of the New York School," *American Journal of Psychiatry* 151 (1994): 482–8.
179 Andreason, *The Creative Brain*, 97–101. Andreason indicates that she herself is embarking on a scientific study of links between scientific ability and

schizophrenia. She expects to find an increased rate of schizophrenia in family members of scientific geniuses (101).

180 On the links between nineteenth-century Gothic romances and science fiction, see Patrick Brantlinger, "The Gothic Origins of Science Fiction," *NOVEL: A Forum on Fiction* 14.1 (1980): 30–43; and Brian Aldiss, *Billion Year Spree: The True History of Science Fiction* (New York: Doubleday, 1973), 18–19.

181 "Who's the Scientist? Seventh-graders describe scientists before and after a visit to Fermilab," June 2010: http://ed.fnal.gov/projects/scientists/index.html (accessed August 3, 2010).

182 See, for instance, Susan Carol Losh, "Stereotypes about Scientists over Time among U.S. Adults: 1983 and 2001," *Public Understanding of Science* 19.3 (May 2010): 372–82; David Wade Chambers, "Stereotypic Images of a Scientist: The Draw-a-Scientist Test," *Science Education* 67.2 (1983): 255–65; and Margaret Mead and R. Metraux, "The Image of the Scientist among High School Students: A Pilot Study," *Science* 126 (1957): 384–90.

## 5 MARIE CORELLI AND THE NEURON

1 Marie Corelli, *Young Diana: An Experiment of the Future* (Whitefish, MT: Kessinger, 1997), 306.

2 Biographer Teresa Ransom writes, "[Corelli's] formal education seems to have been slight and, as was usually the case for Victorian daughters, consisted mainly of music and literature and some languages." *The Mysterious Miss Marie Corelli: Queen of Victorian Bestsellers* (Stroud: Sutton Publishing, 1999), 18.

3 Corelli alluded to the possibility of biological warfare and the atom bomb in *The Secret Power* (1921), while some contend that she predicted X-rays and wireless telegraphy in *A Romance of Two Worlds*. See J. Cuming Walters, "Epilogue" to Bertha Vyver, *Memoirs of Marie Corelli* (London: Alston Rivers Ltd., 1930), 267.

4 Brian Masters, *Now Barabbas Was a Rotter: The Extraordinary Life of Marie Corelli* (London: Hamish Hamilton, 1978), 6.

5 Annette Federico, *Idol of Suburbia: Marie Corelli and Late-Victorian Literary Culture* (Charlottesville: University Press of Virginia, 2000), 2.

6 Federico, *Idol of Suburbia*, 8.

7 Federico, *Idol of Suburbia*, 64–5.

8 Marie Corelli to George Bentley, late March 1887, Corelli Collection (Cat. No. GEN MSS 332), Yale University, Beinecke Rare Book and Manuscript Library. Hereafter referred to as "Corelli Collection."

9 Robyn Hallim, "Marie Corelli's Best-Selling Electric Creed," *Women's Writing* 13.2 (June 2006): 267–83, 268.

10 Masters, *Now Barabbas Was a Rotter*, 94.

11 On Corelli's interest in thermodynamics and electricity, see Hallim's "Electric Creed" and the earlier work of which it is an excerpt, "Marie

Corelli: Science, Society and the Best Seller" (Ph.D. dissertation, University of Sydney, 2002). On Corelli's interest in technologies like the telephone and telegraph, see Jill Galvan, "Christians, Infidels, and Women's Channeling in the Writings of Marie Corelli," *Victorian Literature and Culture* 31.1 (2003): 83–97. On Corelli and psychology, see Alisha Siebers, "Marie Corelli's Magnetic Revitalizing Power," in Martin Willis and Catherine Wynne, eds., *Victorian Literary Mesmerism* (New York: Rodopi, 2006), 183–202; and Alisha Siebers, "The Genius in *Ardath: The Story of A Dead Self*," *Women's Writing* 13.2 (June 2006): 246–66.

12 Like some Victorian spiritualists and heterodox religious sects (such as Christian Scientists), Corelli conceived of God as bisexual, exhibiting both male and female characteristics.

13 Gordon Shepherd, *Foundations of the Neuron Doctrine* (New York: Oxford University Press, 1991), 49–55.

14 The *Journal of Anatomy and Physiology* published a series of articles in 1891 and 1892 introducing central tenets of the neuron doctrine, the first being Albert von Kölliker and William Aldren Turner, "The Minute Anatomy of the Spinal Cord and Cerebellum Demonstrated by Golgi's Method," *Journal of Anatomy and Physiology* 25.3 (1891): 443–60.

15 Stanley Finger, *Minds Behind the Brain: A History of the Pioneers and their Discoveries* (New York: Oxford University Press, 2000), 211–12.

16 Finger, *Minds Behind the Brain*, 197.

17 Stanley Finger, *Origins of Neuroscience: A History of Explorations into Brain Function* (New York: Oxford University Press, 1994), 53; Shepherd, *Foundations*, 259–70.

18 Marie Corelli, *Wormwood* (Whitefish, MT: Kessinger, 2003), 141.

19 There are also explicit references to "brain cells" in chapters 11 and 16 of Oscar Wilde's *Picture of Dorian Gray*, which was written in 1890 and revised in 1891. Since Corelli and Wilde were on friendly terms during this period (that is, before his infamous 1895 trial), it is possible that they discussed brain cells with one another.

20 Marie Corelli, *The Soul of Lilith* (New York: The American News Company, 1892), 34.

21 Marie Corelli, *The Sorrows of Satan* (New York: Oxford University Press, 1996), 304.

22 Robert Louis Stevenson, "A Gossip on Romance," in Stevenson, *The Lantern Bearers and Other Essays*, ed. Jeremy Treglown (London: Chatto & Windus, 1988), 172–82, 175.

23 Marie Corelli, "Author's Prologue" to *The Life Everlasting: A Reality of Romance* (New York: George H. Doran, 1911). Further references to this novel and the "Author's Prologue" that precedes it will be cited parenthetically in the text as *LE*.

24 Marie Corelli, *Free Opinions Freely Expressed* (New York: Dodd, Mead, and Co., 1905), 276.

25 "Appendix" to Marie Corelli, *A Romance of Two Worlds* (New York, A. L. Burt, 1887), 311–24, 314.

26 "Appendix" to Corelli, *A Romance of Two Worlds*, 311–24.
27 "Appendix" to Corelli, *A Romance of Two Worlds*, 311–24.
28 On Corelli's ambivalence toward Victorian gender norms, see Federico, *Idol of Suburbia*, 94–127.
29 On neurasthenia in *A Romance of Two Worlds*, see Siebers, "Marie Corelli's Magnetic Revitalizing Power," 188–90.
30 According to Corelli, the name "Heliobas" is a Chaldean composition of words meaning "beautiful death." It seems worth noting, however, that "Helios" is also the name of the sun god in Greek mythology. Masters, *Now Barabbas Was a Rotter*, 113.
31 Marie Corelli, *A Romance of Two Worlds* (n.p.: Plain Label Books, 1928), 233. Further references to this edition of *Romance* will be cited parenthetically within the text as *RW*.
32 Marie Corelli, "Author's Prologue" to *A Romance of Two Worlds* (1887), 13.
33 Corelli, "Author's Prologue" to *A Romance of Two Worlds*, 16.
34 Marie Corelli to J. Cuming Walters, December 6, 1923. Marie Corelli Papers (Collection 748), Department of Special Collections, Charles E. Young Research Library, UCLA. Hereafter referred to as Corelli Papers.
35 Alex Owen, *The Darkened Room: Women, Power and Spiritualism in Late-Victorian England* (University of Chicago Press, 1989), 1.
36 Janet Oppenheim, *The Other World: Spiritualism and Psychical Research in England, 1850–1914* (Cambridge University Press, 1985), 1–4.
37 Oppenheim, *The Other World*, 50.
38 Oppenheim, *The Other World*, 4.
39 Corelli, "Author's Prologue" to *A Romance of Two Worlds*, 9. On the affinities between Corelli's work and Victorian spiritualism, see Hallim, "Electric Creed" and "Science, Society and the Best Seller"; and Rita Felski, *The Gender of Modernity* (Cambridge, MA: Harvard University Press, 1995), 115–44. On Corelli and mesmerism, see Siebers, "Marie Corelli's Magnetic Revitalizing Power." On Corelli's interest in reincarnation and Victorian popular Buddhism, see J. Jeffrey Franklin, "The Counter-Invasion of Britain by Buddhism in Marie Corelli's *A Romance of Two Worlds* and H. Rider Haggard's *Ayesha: The Return of She*," *Victorian Literature and Culture* 31.1 (2003): 19–42.
40 Marie Corelli, "A Romance of Two Worlds," in *My First Book* (London: Chatto & Windus, 1897), 206–20, 214.
41 Corelli also claimed that her first publisher, George Bentley, had warned that she risked losing her audience if she wrote another "spiritualistic book" (*LE* 22–3).
42 Nearly all Corelli scholarship discusses her strained relationship with literary critics and the press in general. See especially Peter Keating, "Introduction" to Corelli, *The Sorrows of Satan*, ix–xx; Felski, *The Gender of Modernity*, 115–44; and Federico, *Idol of Suburbia*, 14–52.
43 See Charles Mackay, "The Magnetisers," in *Memoirs of Extraordinary Popular Delusions and the Madness of Crowds* (London: Office of the National Illustrated Library, 1841): www.econlib.org/library/mackay/macEx7.html

(July 10, 2008). Mackay referred to himself as Corelli's stepfather, and Corelli lived with him for most of her childhood. In fact, he may have been her natural father. See Masters, *Now Barabbas Was a Rotter*, 18–24, and Ransom, *The Mysterious Miss Marie Corelli*, 225–31.

44 Marie Corelli to George Bentley, December 29, 1888, Corelli Collection.

45 Masters, *Now Barabbas Was a Rotter*, 302, 94.

46 Marie Corelli, *The Mighty Atom* (Philadelphia: J. B. Lippincott, 1896), 4 (emphasis in original).

47 Corelli, *The Mighty Atom*, 4 (emphasis in original).

48 Corelli, "Author's Prologue" to *A Romance of Two Worlds*, 10.

49 Corelli, *Young Diana*, 134. See also pp. 261 and 271.

50 "Our Traditional and Chronological History": www.rosicrucian.org/about/mastery/mastery08history.html (accessed July 7, 2008).

51 Mary Baker Eddy, *Science and Health* (1875) (Boston: The First Church of Christ, Scientist, 1994), 139.

52 Rennie B. Schoepflin, "Christian Science Healing in America," in Norman Gevitz, ed., *Other Healers: Unorthodox Medicine in America* (Baltimore: Johns Hopkins University Press, 1988), 192–214, 199.

53 Marie Corelli to George Bentley, March 5, 1888, Corelli Collection.

54 Marie Corelli to George Bentley, November 30, 1888, Corelli Collection (emphasis in original).

55 "Members and Associates: March, 1907," *Proceedings of the Society for Psychical Research* 19 (1907): 462–4.

56 Frederic Myers, "Introduction" to Edmund Gurney, Frederic Myers, and Frank Podmore, *Phantasms of the Living*, ed. E. M. Sidgwick, 2nd edn. (New York: E. P. Dutton, 1918), xxxiii–lx, lvi. On a tragic side note, Gurney later committed suicide when he discovered that some of the participants in his telepathic experiments had behaved fraudulently, compromising the validity of his findings. Oppenheim, *The Other World*, 144.

57 Roger Luckhurst, *The Invention of Telepathy, 1870–1901* (New York: Oxford University Press, 2002), 75.

58 Oliver Lodge, "An Account of Some Experiments in Thought-Transference," *Proceedings of the Society for Psychical Research* 2 (1884): 189–200, 191.

59 Luckhurst, *The Invention of Telepathy*, 79.

60 Oliver Lodge, *The Survival of Man: A Study in Unrecognized Human Faculty*, 7th edn. (London: Methuen, 1916), 84.

61 William Barrett, "Appendix to the Report on Thought Reading," *Proceedings of the Society for Psychical Research* 1 (1882–3): 47–64, 62.

62 William Barrett, *Nature* (July 7, 1881), 212, quoted in Luckhurst, *The Invention of Telepathy*, 76.

63 Luckhurst, *The Invention of Telepathy*, 82.

64 Lodge, *The Survival of Man*, 80–1.

65 Corelli, quoted in Masters, *Now Barabbas Was a Rotter*, 10.

66 On spiritualists who became disenchanted with the SPR, see Owen, *The Darkened Room*, 102–6.

67 In a letter published posthumously in *The City News* on May 31, 1924, Corelli claims, "I had great friends in the late Sir James Dewar and Sir William Crookes – the latter showing me a private experiment of *working sound from a 'ray' of light*" (emphasis in original). Corelli Papers.
68 Oppenheim, *The Other World*, 386.
69 William Barrett, "The Deeper Issues of Psychical Research," *Contemporary Review* 113 (February 1918): 174, quoted in Oppenheim, *The Other World*, 368.
70 Laura Otis, "The Metaphoric Circuit: Organic and Technological Communication in the Nineteenth Century," *Journal of the History of Ideas* 63.1 (January 2002): 105–28.
71 Sidney Ochs, *A History of Nerve Functions: From Animal Spirits to Molecular Mechanisms* (Cambridge University Press, 2004), 118.
72 Ochs, *History of Nerve Functions*, 118.
73 Anne K. Mellor, *Mary Shelley: Her Life, Her Fiction, Her Monsters* (New York: Methuen, 1988), 104–6.
74 On Galvani's influence on *Frankenstein*, see Mellor, *Mary Shelley*, 89–114; Stanley Finger and Mark B. Law, "Karl August Weinhold and his 'Science' in the Era of Mary Shelley's *Frankenstein*: Experiments on Electricity and the Restoration of Life," *Journal of the History of Medicine* 53.2 (1998): 161–81; and Samuel Holmes Vasbinder, *Scientific Attitudes in Mary Shelley's* Frankenstein (Ann Arbor: University of Michigan Research Press, 1984).
75 Vyver, *Memoirs of Marie Corelli*, 19–34.
76 Johann Wolfgang von Goethe, quoted in Christoph Asendorf, *Batteries of Life: On the History of Things and Their Perception in Modernity*, trans. Don Reneau (Berkeley: University of California Press, 1993), 153.
77 Asendorf, *Batteries of Life*, 153–65.
78 Edwin Clarke and L. S. Jacyna, *Nineteenth-Century Origins of Neuroscientific Concepts* (Berkeley: University of California Press, 1987), 176.
79 Around 1850, the medicinal use of electric fish declined because artificially generated electricity was "more readily available and capable of being better controlled." Ochs, *History of Nerve Functions*, 116.
80 Anson Rabinbach, *The Human Motor: Energy, Fatigue, and the Origins of Modernity* (New York: Basic Books, 1990), 3.
81 Rabinbach, *The Human Motor*, 56.
82 Jacob Moleschott, *Der Kreislauf des Lebens*, 5th edn. (1887), quoted in Rabinbach, *The Human Motor*, 50.
83 Hallim, "Science, Society and the Best Seller," 102.
84 Asendorf, *Batteries of Life*, 160.
85 The destructive potential of electrical technologies became especially apparent in 1889, with the invention of the electric chair. Asendorf, *Batteries of Life*, 163.
86 Clarke and Jacyna, *Nineteenth-Century Origins of Neuroscientific Concepts*, 206.
87 Clarke and Jacyna, *Nineteenth-Century Origins of Neuroscientific Concepts*, 211.

88 Clarke and Jacyna, *Nineteenth-Century Origins of Neuroscientific Concepts*, 261.
89 Robert Young, *Mind, Brain and Adaptation in the Nineteenth Century: Cerebral Localization and Its Biological Context from Gall to Ferrier* (New York: Oxford University Press, 1990), 224–5.
90 Clarke and Jacyna, *Nineteenth-Century Origins of Neuroscientific Concepts*, 183.
91 Carpenter wrote, "Just as a perfectly constructed Galvanic battery is *inactive* while the circuit is 'interrupted,' but becomes *active* the instant that the circuit is 'closed,' so does a Sensation, an Instinctive tendency, an Emotion, an Idea, or a Volition, which attains an intensity adequate to 'close' the circuit, liberate the Nerve-force with which a certain part of the Brain, while in a state of wakeful activity, is always charged" (emphasis in original). William B. Carpenter, *Principles of Mental Physiology* (New York: Appleton, 1874), 14.
92 George Beard, "Preface" to *American Nervousness: Its Causes and Consequences* (New York: G. P. Putnam's Sons, 1881), vi.
93 Beard, *American Nervousness*, 73. On social and scientific factors that contributed to the "invention" of neurasthenia, see Tom Lutz, *American Nervousness, 1903: An Anecdotal History* (Ithaca, NY: Cornell University Press, 1991), 3–7; and Marijke Gijswijt-Hofstra, "Introduction," in Marijke Gijswijt-Hofstra and Roy Porter, eds., *Cultures of Neurasthenia from Beard to the First World War* (New York: Rodopi, 2001), 1–30.
94 Rabinbach, *The Human Motor*, 67.
95 Rabinbach, *The Human Motor*, 164.
96 Beard, *American Nervousness*, 10.
97 Gijswijt-Hofstra, "Introduction," 2.
98 Shepherd, *Foundations*, 275–8.
99 George Beard and A. D. Rockwell, *A Practical Treatise on Nervous Exhaustion (Neurasthenia): Its Symptoms, Nature, Sequences, Treatment* (1905) (New York: Kraus Reprint Co., 1970), 262.
100 Beard and Rockwell, *A Practical Treatise*, 270.
101 Marie Corelli, *Ardath: The Story of a Dead Self* (New York: William Allison, *c.*1900), 26.
102 Corelli, *Soul of Lilith*, 34.
103 Laura Otis, "Howled out of the Country: Wilkie Collins and H. G. Wells Retry David Ferrier," in Anne Stiles, ed., *Neurology and Literature, 1860–1920* (New York: Palgrave, 2007), 27–51, 31. I am also indebted to Paul Brians for pointing out the resonances between *The Matrix* and Corelli's fiction.
104 Corelli, *Young Diana*, 161.
105 Corelli, *Young Diana*, 206.
106 On the electrical properties of Leyden jars, see Ochs, *History of Nerve Functions*, 110–11.
107 Sigmund Freud, *Project for a Scientific Psychology*, in *The Standard Edition of the Complete Psychological Works of Sigmund Freud*, trans. and ed. James Strachey *et al.* (London: Hogarth Press, 1966), 1: 298.

108 Shepherd, *Foundations*, 272–9.
109 Charles Sherrington, *The Integrative Action of the Nervous System* (New York: Charles Scribner's Sons, 1906), 17.
110 Shepherd, *Foundations*, 275.
111 Shepherd, *Foundations*, 279.
112 The dangers of radium were not widely understood and documented until the 1920s, when employees who painted dials and instrument panels with paint containing radium-226 began losing their teeth, developing bone cancers, and exhibiting a range of other ailments. "Radium," *Hazardous Substances Data Bank, National Library of Medicine*: http://toxnet.nlm.nih.gov/cgi-bin/sis/search/r?dbs+hsdb:@term+@na+@rel+radium,+radioactive (accessed February 26, 2007).
113 Marie Corelli, *The Secret Power* (New York: Doubleday Page, 1921), 195–6.
114 Corelli, *The Secret Power*, 268.
115 Ernest Rutherford described the radioactive decay of the element thorium as early as 1900. See Rutherford, "A Radioactive Substance Emitted from Thorium Compounds," *Philosophical Magazine* 49 (1900): 1–14.
116 Marie Curie, "Radium and Radioactivity," *Century Magazine* (January 1904): 461–6.
117 The Roentgen Society of London, *Archives of the Roentgen Ray* (London: Rebman, 1907), 120.

## EPILOGUE: LOOKING FORWARD

1 Patrick Brantlinger, "The Gothic Origins of Science Fiction," *NOVEL: A Forum on Fiction* 14.1 (1980): 30–43; Brian Aldiss, *Billion Year Spree: The True History of Science Fiction* (New York: Doubleday, 1973), 3, 18–19.
2 Oliver Sacks, *The Man Who Mistook His Wife for a Hat and Other Clinical Tales* (New York: Simon & Schuster, 1998), 4–5.
3 Given Crichton's medical training and longstanding interest in meditation, it seems unlikely that the patient's name is an accident. The name "Harold Benson" is strikingly similar to Herbert Benson, a prominent cardiologist who was among the first to document the beneficial health effects of Eastern meditation. While Benson's pathbreaking book, *The Relaxation Response*, did not appear until 1975, he began publishing articles on the healthful effects of yoga in the late 1960s. Benson was one of many mid-century physicians who decried the rapid, stressful pace of modern urban life; his overall message therefore chimes with Crichton's anti-technology stance in *The Terminal Man* (New York: Knopf, 1972).
4 Crichton, *Terminal Man*, 81.
5 George Slusser, "Introduction: Fiction as Information," in George Slusser and Tom Shippey, eds., *Fiction 2000: Cyberpunk and the Future of Narrative* (Athens: University of Georgia Press, 1992), 6.
6 William Gibson, *Neuromancer: 20th Anniversary Edition* (New York: Ace Books, 2004), 7.

7 Some writers are optimistic about the ways in which cyberpunk (and the technologies it imagines) destabilize fundamental categories of understanding. For instance, Donna Haraway argues that "cyborg imagery can suggest a way out of the maze of dualisms in which we have explained our bodies," including seemingly rigid binarisms such as mind/body, human/animal, or male/female. Donna Haraway, *Simians, Cyborgs and Women: The Reinvention of Nature* (New York: Routledge, 1991), 178.
8 Susan Sontag, "The Imagination of Disaster," in *Against Interpretation and Other Essays* (New York: Farrar, Straus, and Giroux, 1967), 207–25, 221–2.
9 See Haraway, *Simians, Cyborgs and Women*, 149–81.
10 Courtney Humphries, "Untangling the Brain: From Neuron to Mind," *Harvard Magazine* 111.5 (2009): 40–6, 42–3.
11 "Harnessing the Power of the Brain," narr. Scott Pelley, *60 Minutes*, CBS, November 2, 2008. Transcript available at: www.cbsnews.com/stories/2008/10/31/60minutes/main4560940_page2.shtml (accessed May 26, 2009).
12 Quoted in "Harnessing the Power of the Brain."
13 On the biological turn in modern psychiatry, see Roy Porter and Mark S. Micale, "Introduction: Reflections on Psychiatry and its Histories," in Micale and Porter, eds., *Discovering the History of Psychiatry* (New York: Oxford University Press, 1994), 3–38, 25; Joseph LeDoux, *Synaptic Self: How Our Brains Become Who We Are* (New York: Penguin, 2002), 260; and Anne Stiles, "Introduction," in Stiles, ed., *Neurology and Literature, 1860–1920* (New York: Palgrave, 2007), 1–23, 19.
14 Some neurologists contest the idea that mental illness is caused by chemical imbalance in the brain. See, for instance, Peter Breggin and Ginger Ross Breggin, *Talking Back to Prozac: What Doctors Won't Tell You about Today's Most Controversial Drug* (New York: St. Martin's Press, 1994), 33.
15 Shari Roan, "Psychiatrists Rewriting the Mental Health Bible," *Los Angeles Times* (May 26, 2009): www.latimes.com/news/nationworld/nation/la-sci-mental-disorder26–2009may26,0,3370054,full.story.
16 To complicate matters further, 56 percent of the individuals currently engaged in revising the *DSM* have ties to the pharmaceutical industry. Roan, "Psychiatrists Rewriting the Mental Health Bible."
17 George Bush, "Presidential Proclamation 6158," July 17, 1990, Library of Congress: www.loc.gov/loc/brain/proclaim.html (accessed May 18, 2009).
18 Humphries, "Untangling the Brain," 46.
19 LeDoux, *Synaptic Self*, 2.
20 LeDoux, *Synaptic Self*, 13.
21 Alan Richardson and Francis F. Steen, "Literature and the Cognitive Revolution: An Introduction," *Poetics Today* 23.1 (2002): 1–8, 3, 2. Richardson maintains a comprehensive bibliography of cognitive literary criticism on his website, "Literature, Cognition, and the Brain": www2.bc.edu/~richarad/lcb/home.html#bib.

22 Mark Turner, "The Cognitive Study of Art, Language, and Literature," *Poetics Today* 23.1 (2002): 9–20, 9.
23 D. T. Max, "The Literary Darwinists," in Atul Gawande, ed., *The Best American Science Writing* (New York: Harper Perennial, 2006), 208–22.
24 For some provisional answers to these questions, see Paul Hernadi, "Why Is Literature: A Coevolutionary Perspective on Imaginative Worldmaking," *Poetics Today* 23.1 (2002): 21–42.
25 The Breggins' *Talking Back to Prozac* is representative of the "Prozac backlash" of the 1990s. For a more recent, even-handed critique of psychiatric drug use, see Ashley Pettus, "Psychiatry by Prescription," *Harvard Magazine* 108.6 (2006): 38–44, 90–9.
26 Quoted in Pettus, "Psychiatry by Prescription": 41.
27 Robert Young, *Mind, Brain and Adaptation in the Nineteenth Century: Cerebral Localization and Its Biological Context from Gall to Ferrier* (New York: Oxford University Press, 1990), x.

# *Bibliography*


"1,000,000 A.D." *Punch* (November 25, 1893): 250

Aldiss, Brian. *Billion Year Spree: The True History of Science Fiction*. New York: Doubleday, 1973

Allen, Grant. *The Colour Sense: Its Origin and Development*. London: Truebner, 1879

"The Dog's Universe." *The Gentleman's Magazine* 249 (1880): 287–301

*The Evolutionist at Large*. London: Chatto & Windus, 1881

*Hilda Wade: A Woman With Tenacity of Purpose*. New York: G. P. Putnam, 1900

"Mr. G. S. Hall on the Perception of Colour." *Mind* 4.14 (April 1879): 267–8

"Personal Reminiscences of Herbert Spencer." *The Forum* 35 (1903–4): 610–28

*Physiological Aesthetics*. New York: D. Appleton and Co., 1877

*Recalled to Life* (1891). Hard Press, 2005

"Review of *L'Hérédité Psychologique*, by Th. Ribot." *Mind* 7.27 (July 1882): 413–15

"Sight and Smell in Vertebrates." *Mind* 6.24 (October 1881): 453–71

*Strange Stories*. London: Chatto & Windus, 1884

"A Thinking Machine." *The Gentleman's Magazine* 260 (1886): 30–41

Andreasen, Nancy. *The Creative Brain: The Science of Genius*. New York: Penguin, 2006

"Creativity and Mental Illness: Prevalence Rates in Writers and Their First-Degree Relatives." *American Journal of Psychiatry* 144 (1987): 1288–92

Anger, Suzy. "Review of '*The Busiest Man in England': Grant Allen and the Writing Trade, 1875–1900*, and: *Grant Allen: Literature and Cultural Politics at the Fin de Siècle*." *Victorian Studies* 49.2 (Winter 2007): 362–5

Asendorf, Christoph. *Batteries of Life: On the History of Things and Their Perception in Modernity*, trans. Don Reneau. Berkeley: University of California Press, 1993

Ashley, Leonard. "Introduction" to Edmund Gurney *et al.*, *Phantasms of the Living*. Gainesville, FL: Scholars' Facsimiles and Reprints, 1970

Atwood, Margaret. "Introduction" to H. G. Wells, *The Island of Dr. Moreau*, ed. Patrick Parrinder. New York: Penguin, 2005. xiii–xxvii

Auerbach, Nina. *Our Vampires, Ourselves*. University of Chicago Press, 1995

Azam, Eugène. *Hypnotisme, Double Conscience, et Altérations de la Personnalité*. Paris: Librairie J. B. Baillière et Fils, 1887

Bain, Alexander. "Education as a Science." *Mind* 2.5 (1877): 1–21

*Mental Science: A Compendium of Psychology, and the History of Philosophy*. New York: D. Appleton, 1886

Baldick, Chris. *In Frankenstein's Shadow: Myth, Monstrosity, and Nineteenth-Century Writing*. Oxford: Clarendon Press, 1987

Barrett, William. "Appendix to the Report on Thought Reading." *Proceedings of the Society for Psychical Research* 1 (1882–3): 47–64

Bazin, André. "The Ontology of the Photographic Image." In Alan Trachtenberg, ed., *Classic Essays on Photography*. New Haven, CT: Leete's Island, 1980. 237–44

Beard, George. *American Nervousness: Its Causes and Consequences*. New York: G. P. Putnam's Sons, 1881

Beard, George and A. D. Rockwell. *A Practical Treatise on Nervous Exhaustion (Neurasthenia): Its Symptoms, Nature, Sequences, Treatment* (1905). New York: Kraus Reprint Co., 1970

Beattie, Hilary. "Father and Son: The Origins of *Strange Case of Dr. Jekyll and Mr. Hyde*." *The Psychoanalytic Study of the Child* 56 (2001): 317–60

Becker, George. *The Mad Genius Controversy: A Study in the Sociology of Deviance*. Beverly Hills: Sage Publications, 1978

Begley, Sharon. "The Sins of the Fathers, Take 2." *Newsweek*. January 17, 2009. www.newsweek.com/2009/01/16/the-sins-of-the-fathers-take-2.html (accessed November 22, 2010)

Belford, Barbara. *Bram Stoker: A Biography of the Author of* Dracula. New York: Knopf, 1996

Benjamin, Walter. "A Short History of Photography" (1931). In Alan Trachtenberg, ed., *Classic Essays on Photography*. New Haven, CT: Leete's Island, 1980. 199–216

Bergonzi, Bernard. *The Early H. G. Wells: A Study of the Scientific Romances*. Manchester University Press, 1961

Block, Ed, Jr. "James Sully, Evolutionary Psychology, and Late-Victorian Gothic Fiction." *Victorian Studies* 25.4 (1982): 443–67

Bogdan, Robert. *Freak Show: Presenting Human Oddities for Amusement and Profit*. University of Chicago Press, 1988

Botting, Fred. *Gothic Romanced: Consumption, Gender, and Technology in Contemporary Fictions*. New York: Routledge, 2008

Bowler, Peter. *The Eclipse of Darwinism: Anti-Darwinian Evolution Theories in the Decades around 1900*. Baltimore: Johns Hopkins University Press, 1983

Brantlinger, Patrick. "The Gothic Origins of Science Fiction." *NOVEL: A Forum on Fiction* 14.1 (1980): 30–43

*Rule of Darkness: British Literature and Imperialism, 1830–1914*. Ithaca, NY: Cornell University Press, 1988

Breggin, Peter and Ginger Ross Breggin. *Talking Back to Prozac: What Doctors Won't Tell You about Today's Most Controversial Drug*. New York: St. Martin's Press, 1994

Breuer, Josef and Sigmund Freud. *Studies on Hysteria*, trans. James Strachey. New York: Basic Books, 1895

Bruhm, Steven. *Gothic Bodies: The Politics of Pain in Romantic Fiction*. Philadelphia: University of Pennsylvania Press, 1994

Burwick, Frederick. *Poetic Madness and the Romantic Imagination*. University Park: Pennsylvania State University Press, 1996

Bush, George. "Presidential Proclamation 6158." July 17, 1990. Library of Congress. www.loc.gov/loc/brain/proclaim.html (accessed May 18, 2008)

Campion-Vincent, Véronique. "The Tell-Tale Eye." *Folklore* 110 (1999): 13–24

Carpenter, William Benjamin. "On the Doctrine of Human Automatism." *Contemporary Review* 25 (1875): 397–416

"On the Doctrine of Human Automatism: Part II." *Contemporary Review* 25 (1875): 940–62

*Principles of Mental Physiology*. New York: Appleton, 1874

Carroy, Jacqueline. "Entre mémoire et oubli: les deux vies de Félida." *Revue internationale de psychopathologie* 5 (1992): 73–80

*Hypnose, suggestion et psychologie. L'invention des sujets*. Paris: Presses Universitaires de France, 1991

"Le savant et ses doubles. Docteurs Jekyll et Misters Hyde du siècle dernier." *Frénesie* 10 (Spring 1992): 125–42

Chambers, David Wade. "Stereotypic Images of a Scientist: The Draw-a-Scientist Test." *Science Education* 67.2 (1983): 255–65

Clarke, Edwin and L. S. Jacyna. *Nineteenth-Century Origins of Neuroscientific Concepts*. Berkeley: University of California Press, 1987

Clausson, Nils. "Degeneration, *Fin-de-Siècle* Gothic, and the Science of Detection: Arthur Conan Doyle's *The Hound of the Baskervilles* and the Emergence of the Modern Detective Story." *Journal of Narrative Theory* 35.1 (Winter 2005): 60–87

Clifford, William Kingdon. "Body and Mind." In *Lectures and Essays by the Late William Kingdon Clifford, F.R.S.*, ed. Leslie Stephen and Fredrick Pollock. London: Macmillan, 1879

Cobbe, Frances Power. *The Modern Rack: Papers on Vivisection*. London: Swan Sonnenschein, 1889

Coe, Brian. "Eadweard Muybridge." In Stephen Herbert and Luke McKernan, eds., *Who's Who of Victorian Cinema*. London: British Film Institute, 1996. 99–100

Conan Doyle, Arthur. "A Scandal in Bohemia." In *The New Annotated Sherlock Holmes*, ed. Leslie Klinger. New York: W. W. Norton and Company, 2005

*A Study in Scarlet*. In *A Study in Scarlet and Sign of the Four*. New York: Berkley, 1963

Corelli, Marie. *Ardath: The Story of a Dead Self*. New York: William Allison, *c*.1900

*Free Opinions Freely Expressed*. New York: Dodd, Mead, and Co., 1905

*The Life Everlasting: A Reality of Romance*. New York: George H. Doran, 1911

*The Mighty Atom*. Philadelphia: J. B. Lippincott, 1896
*A Romance of Two Worlds*. New York: A. L. Burt, 1887
*A Romance of Two Worlds*. N.p.: Plain Label Books, 1928
"A Romance of Two Worlds." In *My First Book*. London: Chatto & Windus, 1897. 206–20
*The Secret Power*. New York: Doubleday Page, 1921
*The Sorrows of Satan*. New York: Oxford University Press, 1996
*The Soul of Lilith*. New York: The American News Company, 1892
*Wormwood*. Whitefish, MT: Kessinger, 2003
*Young Diana: An Experiment of the Future*. Whitefish, MT: Kessinger, 1997

Corelli, Marie *et al*. Corelli Collection (Cat. No. GEN MSS 332), Yale University, Beinecke Rare Book and Manuscript Library
Marie Corelli Papers (Collection 748), Department of Special Collections, Charles E. Young Research Library, UCLA

Crichton, Michael. *The Terminal Man*. New York: Knopf, 1972

Curie, Marie. "Radium and Radioactivity." *Century Magazine* (January 1904): 461–6

Daly, Nicholas. *Modernism, Romance, and the* Fin de Siècle: *Popular Fiction and British Culture, 1880–1914*. New York: Cambridge University Press, 1999

Dames, Nicholas. *Amnesiac Selves: Nostalgia, Forgetting, and British Fiction, 1810–1870*. New York: Oxford University Press, 2001
*The Physiology of the Novel: Reading, Neural Science, and the Form of Victorian Fiction*. New York: Oxford University Press, 2007

Darwin, Charles *et al*. *Darwin*, ed. Philip Appleman. New York: W. W. Norton and Company, 2001

DeLamotte, Eugenia. *Perils of the Night: A Feminist Study of Nineteenth-Century Gothic*. New York: Oxford University Press, 1990

Descartes, René. *The World and Other Writings*, ed. Stephen Gaukroger. Cambridge University Press, 1998

Dowling, John E. *Neurons and Networks: An Introduction to Behavioral Neuroscience*. Cambridge, MA: Harvard University Press, 2001

Draaisma, Douwe. *Metaphors of Memory: A History of Ideas about the Mind*, trans. Paul Vincent. Cambridge University Press, 2000

Drinka, George Frederick. *The Birth of Neurosis: Myth, Malady, and the Victorians*. New York: Simon & Schuster, 1984

Duncan, Ian. *Modern Romance and Transformations of the Novel: The Gothic, Scott, Dickens*. New York: Cambridge University Press, 1992

Dury, Richard. "Crossing the Bounds of Single Identity: *Dr. Jekyll and Mr. Hyde* and a Paper in a French Scientific Journal." In Richard Ambrosini and Richard Dury, eds., *Robert Louis Stevenson, Writer of Boundaries*. Madison: University of Wisconsin Press, 2006. 237–51

Eddy, Mary Baker. *Science and Health*. Boston: The First Church of Christ, Scientist, 1994

Falconer, Paul. "Muybridge's Window to the Past: A Wet-plate View of San Francisco in 1877." *California History* 57 (1978): 130–57

Federico, Annette. *Idol of Suburbia: Marie Corelli and Late-Victorian Literary Culture*. Charlottesville: University Press of Virginia, 2000

Felski, Rita. *The Gender of Modernity*. Cambridge, MA: Harvard University Press, 1995

Ferrier, David. "The Brain of a Criminal Lunatic." *Brain: A Journal of Neurology* 5 (April 1882): 62–73

*The Functions of the Brain*. New York: G. P. Putnam's Sons, 1876

*The Localization of Cerebral Disease*. New York: G. P. Putnam's Sons, 1879

Finger, Stanley. *Minds Behind the Brain: A History of the Pioneers and Their Discoveries*. New York: Oxford University Press, 2000

*Origins of Neuroscience: A History of Explorations into Brain Function*. New York: Oxford University Press, 1994

Finger, Stanley, Sarah Gehr, and Alison West. "'Dual Brain Action': The Case Studies of Lewis C. Bruce in the 1890s." *History of Psychiatry* 12 (2001): 59–71

Finger, Stanley and Mark B. Law. "Karl August Weinhold and His 'Science' in the Era of Mary Shelley's *Frankenstein*: Experiments on Electricity and the Restoration of Life." *Journal of the History of Medicine* 53.2 (1998): 161–81

Fleetwood, John. *History of Medicine in Ireland*. Dublin: The Richview Press, 1951

Fleissner, Jennifer. "Dictation Anxiety: The Stenographer's Stake in Dracula." *Nineteenth-Century Contexts* 22 (2000): 417–55

Flint, Kate. "Throwing Light on the Moment: Flash Photography and Arresting Narratives." Paper presented at the 6th annual North American Victorian Studies Association Conference, New Haven, CT, November 14–16, 2008

Flood, David. "Blood and Transfusion in Bram Stoker's *Dracula*." *University of Mississippi Studies in English*, New Series, 7 (1989): 180–92

Fontana, Ernest. "Lombroso's Criminal Man and Stoker's *Dracula*." *Victorian Newsletter* 66 (1984): 25–7

Foucault, Michel. *The Birth of the Clinic: An Archaeology of Medical Perception*, trans. A. M. Sheridan Smith. New York: Vintage Books, 1994

*The History of Sexuality, Volume One: An Introduction*, trans. Robert Hurley. New York: Vintage Books, 1978

Frank, Lawrence. *Victorian Detective Fiction and the Nature of Evidence: The Scientific Investigations of Poe, Dickens, and Doyle*. New York: Palgrave, 2003

Franklin, J. Jeffrey. "The Counter-Invasion of Britain by Buddhism in Marie Corelli's *A Romance of Two Worlds* and H. Rider Haggard's *Ayesha: The Return of She*." *Victorian Literature and Culture* 31.1 (2003): 19–42

French, Richard. *Antivivisection and Medical Science in Victorian Society*. Princeton University Press, 1975

Freud, Sigmund. *The Standard Edition of the Complete Psychological Works of Sigmund Freud*, trans. and ed. James Strachey *et al*. London: Hogarth Press, 1966

Furst, Lilian. *Between Doctors and Patients: The Changing Balance of Power*. Charlottesville: University Press of Virginia, 1998

Galton, Francis. *English Men of Science: Their Nature and Nurture*. London: Macmillan and Co., 1874

*Hereditary Genius: An Inquiry into Its Laws and Consequences*. London: Macmillan and Co., 1869

*Hereditary Genius: An Inquiry into Its Laws and Consequences*. London: Watts and Co., 1950

Galvan, Jill. "Christians, Infidels, and Women's Channeling in the Writings of Marie Corelli." *Victorian Literature and Culture* 31.1 (2003): 83–97

Garrett, Peter. *Gothic Reflections: Narrative Force in Nineteenth-Century Fiction*. Ithaca, NY: Cornell University Press, 2003

Gibson, William. *Neuromancer: 20th Anniversary Edition*. New York: Ace Books, 2004

Gijswijt-Hofstra, Marijke and Roy Porter, eds. *Cultures of Neurasthenia from Beard to the First World War*. New York: Rodopi, 2001

Glover, David. *Vampires, Mummies and Liberals: Bram Stoker and the Politics of Popular Fiction*. Durham, NC: Duke University Press, 1996

Goldstein, Jan. *Console and Classify: The French Psychiatric Profession in the Nineteenth Century*. Cambridge University Press, 1987

"The Uses of Male Hysteria: Medical and Literary Discourse in Nineteenth-Century France." *Representations* 34 (1991): 134–65

Gould, Steven J. *The Mismeasure of Man*. New York: W. W. Norton and Company, 1981

Goulet, Andrea. "Retinal Fictions: Villiers, Leroux, and Optics at the Fin-de-Siècle." *Nineteenth-Century French Studies* 34.1–2 (2005–6): 107–20

Gray, B. M. "Pseudoscience and George Eliot's 'The Lifted Veil.'" *Nineteenth-Century Fiction* 36.4 (1982): 407–23

Gray, Robert. *The Theory of Dreams: In Which an Inquiry Is Made into the Powers and Faculties of the Human Mind, as They Are Illustrated in the Most Remarkable Dreams*. 2 vols. London: F. C. and J. Livingston, 1808

Glendening, John. "'Green Confusion': Evolution and Entanglement in H. G. Wells's *Island of Doctor Moreau*." *Victorian Literature and Culture* 30.2 (2002): 571–97

Greenslade, William. *Degeneration, Culture, and the Novel 1880–1940*. Cambridge University Press, 1994

Greenway, John L. "Seward's Folly: *Dracula* as a Critique of 'Normal Science.'" *Stanford Literature Review* 3.2 (1986): 213–30

"'Unconscious Cerebration' and the Happy Ending of *Dracula*." *Journal of Dracula Studies* 4 (2002): 1–9

Gurney, Edmund, Frederic Myers, and Frank Podmore. *Phantasms of the Living*, ed. E. M. Sidgwick. 2nd edn. New York: E. P. Dutton, 1918

Hacking, Ian. "*Automatisme Ambulatoire*: Fugue, Hysteria, and Gender at the Turn of the Century." *Modernism/Modernity* 3.2 (April 1996): 31–43

*Rewriting the Soul: Multiple Personality and the Sciences of Memory*. Princeton University Press, 1995

"Two Souls in One Body." *Critical Inquiry* 17.4 (Summer 1991): 838–67

Haight, Gordon S. "H. G. Wells's 'Man of the Year Million.'" *Nineteenth-Century Fiction* 12.4 (March 1958): 323–6

Hall, G. Stanley and E. M. Hartwell. "Bilateral Asymmetry of Function." *Mind* 9.33 (January 1884): 94–109

Hallim, Robyn. "Marie Corelli: Science, Society and the Best Seller." Ph.D. dissertation, University of Sydney, 2002

"Marie Corelli's Best-selling Electric Creed." *Women's Writing* 13.2 (2006): 267–83

Haraway, Donna. *Simians, Cyborgs and Women: The Reinvention of Nature*. New York: Routledge, 1991

"Harnessing the Power of the Brain." Narr. Scott Pelley. *60 Minutes*. CBS. November 2, 2008

Harrington, Anne. *Medicine, Mind, and the Double Brain: A Study in Nineteenth-Century Thought*. Princeton University Press, 1987

Harrison, Peter. "Descartes on Animals." *The Philosophical Quarterly* 42.167 (1992): 219–27

Haynes, Renée. *The Society for Psychical Research, 1882–1982: A History*. London: Macdonald, 1982

Haynes, Roslynn D. *From Faust to Strangelove: Representations of the Scientist in Western Literature*. Baltimore: Johns Hopkins University Press, 1994

*H. G. Wells: Discoverer of the Future – The Influence of Science on His Thought*. London: Macmillan, 1980

"Wells's Debt to Huxley and the Myth of Dr. Moreau." *Cahiers Victoriens et Edouardiens* 13 (1981): 31–41

Heath, Stephen. "Psychopathia Sexualis: Stevenson's *Strange Case*." *Critical Quarterly* 28.1 (1986): 93–108

Hendricks, Gordon. *Eadweard Muybridge: The Father of the Motion Picture*. New York: Viking, 1975

Hernadi, Paul. "Why Is Literature: A Coevolutionary Perspective on Imaginative Worldmaking." *Poetics Today* 23.1 (2002): 21–42

Holmes, Oliver Wendell, Sr. "Crime and Automatism." *Atlantic Monthly* 35 (1875): 466–81

"The Stereoscope and the Stereograph" (1859). In Alan Trachtenberg, ed., *Classic Essays on Photography*. New Haven, CT: Leete's Island, 1980

"How to Humanize the Landscape." *Punch* (March 15, 1911): 195

Hughes, David Y. "A Queer Notion of Grant Allen's." *Science Fiction Studies* 25.2 (June 1998): 271–84

Hughes, William. *Beyond Dracula: Bram Stoker's Fiction and Its Cultural Context*. New York: Macmillan, 2000

*Bram Stoker:* Dracula, *A Reader's Guide to Essential Criticism*. New York: Palgrave, 2009

"'So Unlike the Normal Lunatic': Abnormal Psychology in Bram Stoker's *Dracula*." *University of Mississippi Studies in English* 11/12 (1993–5): 1–10

Humphries, Courtney. "Untangling the Brain: From Neuron to Mind." *Harvard Magazine* 111.5 (2009): 40–6

Hungerford, Edward. "Poe and Phrenology." *American Literature* 2.3 (November 1930): 209–31

Hurley, Kelly. *The Gothic Body: Sexuality, Materialism, and Degeneration at the* Fin de Siècle. New York: Cambridge University Press, 1996

Hutcheon, Linda. *The Politics of Postmodernism*. 2nd edn. New York: Routledge, 2002

Huxley, Thomas Henry. *Evolution and Ethics*. New York: Prometheus Books, 2004

*Method and Results: Essays by Thomas Henry Huxley*. New York: D. Applegate and Co., 1901

James, Henry and Robert Louis Stevenson. *Henry James and Robert Louis Stevenson: A Record of Friendship and Criticism*, ed. Janet Adam Smith. London: Rupert Hart-Davis, 1948

James, William. *Essays in Psychical Research*. Cambridge, MA: Harvard University Press, 1986

*The Principles of Psychology*. 2 vols. New York: Henry Holt, 1923

Jamison, Kay Redfield. "Mood Disorders and Patterns of Creativity in British Writers and Artists." *Psychiatry* 52 (1989): 125–34

Jenkins, Alice. *Space and the "March of Mind": Literature and the Physical Sciences in Britain 1815–1850*. New York: Oxford University Press, 2007

Keating, Peter. "Introduction" to Marie Corelli, *The Sorrows of Satan*. New York: Oxford University Press, 1996

Kemp, Peter. *H. G. Wells and the Culminating Ape*. New York: St. Martin's, 1982

Kestner, Joseph. *Sherlock's Sisters: The British Female Detective, 1864–1913*. Burlington, VT: Ashgate, 2003

Kölliker, Albert von and William Aldren Turner. "The Minute Anatomy of the Spinal Cord and Cerebellum Demonstrated by Golgi's Method." *Journal of Anatomy and Physiology* 25.3 (1891): 443–60

Lamarck, Jean Baptiste. *Zoological Philosophy: An Exposition With Regard to the Natural History of Animals*. University of Chicago Press, 1984

Lange-Eichbaum, Wilhelm. *Genie, Irrsinn und Ruhm*. 7th edn. 12 vols. Munich: Ernst Reinhardt Verlag, 1986

Lankester, E. Ray. *Degeneration. A Chapter in Darwinism*. London: Macmillan and Co., 1880

Lansbury, Coral. *The Old Brown Dog: Women, Workers and Vivisection in Edwardian England*. Madison: University of Wisconsin Press, 1985

LeDoux, Joseph. *Synaptic Self: How Our Brains Become Who We Are*. New York: Penguin, 2002

Lee, Sarah. *Anecdotes on the Habits and Instincts of Animals*. New York: E. P. Dutton, 1852

Le Gallienne, Richard. *Attitudes and Avowals, With Some Retrospective Reviews*. London: John Lane, 1910

Levine, George. *The Realistic Imagination: English Fiction from Frankenstein to Lady Chatterley*. University of Chicago Press, 1981

Lightman, Bernard. *Victorian Popularizers of Science, Designing Nature for New Audiences*. University of Chicago Press, 2007

Lodge, Oliver. "An Account of Some Experiments in Thought-Transference." *Proceedings of the Society for Psychical Research* 2 (1884): 189–200

*The Survival of Man: A Study in Unrecognized Human Faculty*. 7th edn. London: Methuen, 1916

Loentz, Elizabeth. *Let Me Continue to Speak the Truth: Bertha Pappenheim as Author and Activist*. Cincinnati, OH: Hebrew Union College Press, 2007

Lombroso, Cesare. *The Man of Genius*. New York: Garland Publishing, 1984

Losh, Susan Carol. "Stereotypes about Scientists over Time among U.S. Adults: 1983 and 2001." *Public Understanding of Science* 19.3 (May 2010): 372–82

Luce, J. V. *Trinity College Dublin: The First 400 Years*. Trinity College Dublin Press, 1992

Luckhurst, Roger. *The Invention of Telepathy, 1870–1901*. New York: Oxford University Press, 2002

Luria, A. R. *The Mind of a Mnemonist*, trans. Lynn Solotaroff. New York: Basic Books, 1968

Lutz, Tom. *American Nervousness, 1903: An Anecdotal History*. Ithaca, NY: Cornell University Press, 1991

McGurl, Mark. "Zombie Renaissance: Eating Your Brains." *N+1* (April 27, 2010): http://nplusonemag.com/the-zombie-renaissance-r-n (accessed October 25, 2010)

Mackay, Charles. "The Magnetisers." In *Memoirs of Extraordinary Popular Delusions and the Madness of Crowds*. London: Office of the National Illustrated Library, 1841. www.econlib.org/library/mackay/macEx7.html (accessed July 10, 2008)

McLean, Steven. *The Early Fiction of H. G. Wells: Fantasies of Science*. New York: Palgrave, 2009

Maixner, Paul, ed. *Robert Louis Stevenson: The Critical Heritage*. Boston: Routledge, 1981

Manlove, Colin. "Charles Kingsley, H. G. Wells, and the Machine in Victorian Fiction." *Nineteenth-Century Literature* 48.2 (September 1993): 212–39

Masters, Brian. *Now Barabbas Was a Rotter: The Extraordinary Life of Marie Corelli*. London: Hamish Hamilton, 1978

Matus, Jill. "Emergent Theories of Victorian Mind Shock: From War and Railway Accident to Nerves, Electricity and Emotion." In Anne Stiles, ed., *Neurology and Literature, 1860–1920*. New York: Palgrave, 2007. 163–83

"Trauma, Memory, and Railway Disaster: The Dickensian Connection." *Victorian Studies* 43.3 (2001): 413–36

Maudsley, Henry. *The Physiology of Mind*. New York: D. Appleton, 1878

*The Physiology of Mind*. New York: D. Appleton, 1889

"Sex in Mind and Education." *Fortnightly Review* 15 (1874): 469–71

Max, D. T. "The Literary Darwinists." In Atul Gawande, ed., *The Best American Science Writing*. New York: Harper Perennial, 2006. 208–22

Mead, Margaret and R. Metraux. "The Image of the Scientist among High School Students: A Pilot Study." *Science* 126 (1957): 384–90

Mellor, Anne K. *Mary Shelley: Her Life, Her Fiction, Her Monsters*. New York: Methuen, 1988

"Members and Associates." *Proceedings of the Society for Psychical Research* 19 (1907): 462–4

Mesnet, Ernest. *De L'Automatisme de la Mémoire et du Souvenir dans le Somnambulisme Pathologique*. Paris: Imprimerie de Félix Malteste, 1874

Meyer, Stephenie. *New Moon*. New York: Little, Brown and Company, 2006

Micale, Mark. "Charcot and the Idea of Hysteria in the Male: Gender, Mental Science, and Medical Diagnosis in Late Nineteenth-Century France." *Medical History* 34 (1990): 363–411

*Hysterical Men: The Hidden History of Male Nervous Illness*. Cambridge, MA: Harvard University Press, 2008

Mighall, Robert. *A Geography of Victorian Gothic Fiction: Mapping History's Nightmares*. New York: Oxford University Press, 1999

Miller, Elizabeth Carolyn. *Framed: The New Woman Criminal in British Culture at the* Fin de Siècle. Ann Arbor: University of Michigan Press, 2008

Miller, J. Hillis. *The Disappearance of God: Five Nineteenth-Century Writers*. Champaign, IL: University of Illinois Press, 2000

Miyoshi, Masao. "Dr. Jekyll and the Emergence of Mr. Hyde." *College English* 27.6 (1966): 470–80

Moreau, Jacques. *La psychologie morbide dans ses rapports avec la philosophie de l'histoire ou de l'influence des névropathies sur le dynamisme intellectual*. Paris: Librairie Victor Masson, 1859

Morton, Peter. *"The Busiest Man in England": Grant Allen and the Writing Trade, 1875–1900*. New York: Palgrave, 2005

Mulvey-Roberts, Marie. "*Dracula* and the Doctors: Bad Blood, Menstrual Taboo and the New Woman." In William Hughes and Andrew Smith, eds., *Bram Stoker: History, Psychoanalysis and the Gothic*. New York: Macmillan, 1998. 78–95

Murray, Paul. *From the Shadow of* Dracula: *A Life of Bram Stoker*. London: Jonathan Cape, 2004

Myers, Frederic W. H. *Human Personality and Its Survival of Bodily Death*, abridged edn., ed. Leopold Hamilton Myers. London: Longmans, Green, and Co., 1918

"Multiplex Personality." *The Nineteenth Century* 20 (November 1886): 648–66

Naugrette, Jean-Pierre. "L'Étrange case du doutor Pereira et du docteur Cardoso: essai sur la fonction cognitive et politique d'un mythe littéraire." *La Licorne* 55 (2000): 277–92

"Le texte et son double: le cas de M.P. et du Dr. Forsyth." *Otrante* 8 (1995–6): 149–59

Nettle, Daniel. *Strong Imagination: Madness, Creativity and Human Nature.* New York: Oxford University Press, 2001

Nisbet, John Ferguson. *The Insanity of Genius and the General Inequality of Human Faculty Physiologically Considered.* 6th edn. New York: Charles Scribner, 1912

Nordau, Max. *Degeneration*, trans. George L. Mosse. Lincoln: University of Nebraska Press, 1993

Novak, Daniel. *Realism, Photography, and Nineteenth-Century Fiction.* Cambridge University Press, 2008

"Obituary: Sir William Thornley Stoker, Bart., M.D." *British Medical Journal* (June 15, 1912): 1399–400

Ochs, Sidney. *A History of Nerve Functions: From Animal Spirits to Molecular Mechanisms.* Cambridge University Press, 2004

Oppenheim, Janet. *The Other World: Spiritualism and Psychical Research in England, 1850–1914.* Cambridge University Press, 1985

Otis, Laura. "Howled out of the Country: Wilkie Collins and H. G. Wells Retry David Ferrier." In Anne Stiles, ed., *Neurology and Literature, 1860–1920.* New York: Palgrave, 2007. 27–51

"The Metaphoric Circuit: Organic and Technological Communication in the Nineteenth Century." *Journal of the History of Ideas* 63.1 (January 2002): 105–28

*Networking: Communicating With Bodies and Machines in the Nineteenth Century.* Ann Arbor: University of Michigan Press, 2001

Owen, Alex. *The Darkened Room: Women, Power and Spiritualism in Late-Victorian England.* University of Chicago Press, 1989

Pettus, Ashley. "Psychiatry by Prescription." *Harvard Magazine* 108.6 (2006): 38–44 and 90–9

Parrinder, Patrick. "How Far Can We Trust the Narrator of *The War of the Worlds*?" *Foundation* 77 (1999): 15–24

Pick, Daniel. *Faces of Degeneration: A European Disorder, c.1848–1918.* Cambridge University Press, 1989

"'Terrors of the Night': *Dracula* and 'Degeneration' in the Late Nineteenth Century." *Critical Quarterly* 30 (1988): 71–87

Porter, Roy and Mark S. Micale. "Introduction: Reflections on Psychiatry and Its Histories." In Roy Micale and Mark S. Porter, eds., *Discovering the History of Psychiatry.* New York: Oxford University Press, 1994. 3–38

Proctor, Richard. "Dual Consciousness." *The Cornhill Magazine* 35 (January 1877): 86–105

"Have We Two Brains?" *The Cornhill Magazine* 31 (1875): 149–66

Prodger, Phillip. *Muybridge and the Instantaneous Photography Movement.* New York: Oxford University Press in association with the Iris and B. Gerald Cantor Center for Visual Arts at Stanford University, 2003

Punter, David. *The Literature of Terror: A History of Gothic Fiction from 1765 to the Present Day.* London: Longman, 1980

Rabinbach, Anson. *The Human Motor: Energy, Fatigue, and the Origins of Modernity*. New York: Basic Books, 1990

Ransom, Teresa. *The Mysterious Miss Marie Corelli: Queen of Victorian Bestsellers*. Stroud: Sutton Publishing, 1999

Reid, Julia. *Robert Louis Stevenson, Science, and the* Fin de Siècle. New York: Palgrave, 2006

"Review of *Difference in the Nervous Organisation of Man and Woman*, by Harry Campbell." *Mind* 16.63 (1891): 542

"Review of *The Insanity of Genius and the General Inequality of Human Faculty, Physiologically Considered*, by J. F. Nisbet." *Mind* 16 (1891): 541

Ribot, Théodule. *Diseases of Memory: An Essay in the Positive Psychology*, 5th edn. London: Kegan Paul, 1906

Richardson, Alan. *British Romanticism and the Science of the Mind*. Cambridge University Press, 2001

"Literature, Cognition, and the Brain." www2.bc.edu/~richarad/lcb/home.html#bib (accessed July 22, 2009)

Richardson, Alan and Francis F. Steen. "Literature and the Cognitive Revolution: An Introduction." *Poetics Today* 23.1 (2002): 1–8

Roan, Shari. "Psychiatrists Rewriting the Mental Health Bible." *Los Angeles Times* (May 26, 2009): www.latimes.com/news/nationworld/nation/la-sci-mental-disorder26–2009may26,0,3370054,full.story

The Roentgen Society of London. *Archives of the Roentgen Ray*. London: Rebman, 1907

Rogers, Kara. "The Rebirth of Lamarckism (the Rise of Epigenetics)." *Encyclopaedia Britannica Blog* (February 25, 2009): www.britannica.com/blogs/2009/02/the-rebirth-of-lamarckism-the-rise-of-epigenetics (accessed November 22, 2010)

Rothfield, Lawrence. *Vital Signs: Medical Realism in Nineteenth-Century Fiction*. Princeton University Press, 1992

"Royal Commission on Vivisection: Inspection in Ireland." *British Medical Journal* (March 2, 1907): 507–12

Rozendal, Phyllis. "Grant Allen." In *British Mystery Writers, 1860–1919*, vol. LXX of Bernard Benstock and Thomas F. Staley, eds., *Dictionary of Literary Biography*. Detroit: Gale, 1988. 3–13

Rutherford, Ernest. "A Radioactive Substance Emitted from Thorium Compounds." *Philosophical Magazine* 49 (1900): 1–14

Rylance, Rick. *Victorian Psychology and British Culture, 1850–1880*. New York: Oxford University Press, 2000

Sacks, Oliver. *The Man Who Mistook His Wife for a Hat and Other Clinical Tales*. New York: Simon & Schuster, 1998

Saunders, Corinne and Jane Macnaughton, eds. *Madness and Creativity in Literature and Culture*. New York: Palgrave, 2005

Schildkraut, Joseph *et al.* "Mind and Mood in Modern Art, II: Depressive Disorders, Spirituality, and Early Deaths in the Abstract Expressionist

Artists of the New York School." *American Journal of Psychiatry* 151 (1994): 482–8

Schoepflin, Rennie B. "Christian Science Healing in America." In Norman Gevitz, ed., *Other Healers: Unorthodox Medicine in America*. Baltimore: Johns Hopkins University Press, 1988. 192–214

Searleman, Alan. "Is There Such a Thing as a Photographic Memory? And If So, Can It Be Learned?" *Scientific American* (March 12, 2007): www.sciam.com/article.cfm?id=is-there-such-a-thing-as (accessed November 2, 2008)

Senf, Carol. *Science and Social Science in Bram Stoker's Fiction*. Westport, CT: Greenwood Press, 2002

Shepherd, Gordon. *Foundations of the Neuron Doctrine*. New York: Oxford University Press, 1991

Sherrington, Charles. *The Integrative Action of the Nervous System*. New York: Charles Scribner's Sons, 1906

Showalter, Elaine. "Dr. Jekyll's Closet." In Elton Smith and Robert Haas, eds., *The Haunted Mind: The Supernatural in Victorian Literature*. London: Scarecrow Press, 1999

*The Female Malady: Women, Madness, and English Culture, 1830–1980*. New York: Pantheon, 1985

*Sexual Anarchy: Gender and Culture at the* Fin de Siècle. New York: Penguin, 1990

Siebers, Alisha. "The Genius in *Ardath: The Story of A Dead Self*." *Women's Writing* 13.2 (June 2006): 246–66

"Marie Corelli's Magnetic Revitalizing Power." In *Victorian Literary Mesmerism*, ed. Martin Willis and Catherine Wynne. New York: Rodopi, 2006. 183–202

Sleigh, Charlotte. "Empire of the Ants: H. G. Wells and Tropical Entomology." *Science as Culture* 10 (2001): 33–71

Slusser, George. "Introduction: Fiction as Information." In George Slusser and Tom Shippey, eds., *Fiction 2000: Cyberpunk and the Future of Narrative*. Athens: University of Georgia Press, 1992

Small, Helen. "Introduction" to George Eliot, *The Lifted Veil and Brother Jacob*. New York: Oxford University Press, 1999. ix–xxxviii

Smith, Janet Adam, ed. *Henry James and Robert Louis Stevenson: A Record of Friendship and Criticism*. London: R. Hart Davis, 1948

Sontag, Susan. "The Imagination of Disaster." In *Against Interpretation and Other Essays*. New York: Farrar, Straus, and Giroux, 1967. 207–25

Spencer, Herbert. *The Principles of Psychology*, 2 vols. New York: D. Appleton, 1897

Springer, Sally and Georg Deutsch. *Left Brain/Right Brain: Perspectives from Cognitive Neuroscience*, 5th edn. New York: W. H. Freeman, 1997

Stevenson, Fanny Osborne. "Prefatory Note" to Robert Louis Stevenson, *The Strange Case of Dr. Jekyll and Mr. Hyde; Fables, Other Stories and Fragments*. London: Heinemann, 1924. xv–xvi

Stevenson, Robert Louis. Interview. "Mr. R. L. Stevenson on Reading and Literature. An Interview." *The Argus* (April 11, 1893)

*The Lantern Bearers and Other Essays*, ed. Jeremy Treglown. London: Chatto & Windus, 1988

"My First Book" (1894). In *Treasure Island*, ed. Emma Letley. New York: Oxford University Press, 1985. 192–200

*R. L. Stevenson on Fiction: An Anthology of Literary and Critical Essays*, ed. Glenda Norquay. Edinburgh University Press, 1999

*Robert Louis Stevenson: The Scottish Stories and Essays*, ed. Kenneth Gelder. Edinburgh University Press, 1989

*Strange Case of Dr. Jekyll and Mr. Hyde*, ed. Katherine Linehan. New York: W. W. Norton and Company, 2003

*The Strange Case of Dr. Jekyll and Mr. Hyde*, 2nd edn., ed. Martin Danahay. Peterborough, Ontario: Broadview, 2005

Stiles, Anne. "Introduction" to *Neurology and Literature, 1860–1920*, ed. Anne Stiles. New York: Palgrave, 2007. 1–23

"Victorian Psychology and the Novel." *Literature Compass* 5.3 (2008): 668–80

Stiles, Anne, Stanley Finger, and John Bulevich. "Somnambulism and Trance States in the Works of John William Polidori, Author of *The Vampyre*." *European Romantic Review* 21.6 (2010): 789–807

Stoker, Bram. *Dracula*, ed. Nina Auerbach and David Skal. New York: W. W. Norton and Company, 1997

*Personal Reminiscences of Henry Irving*. London: William Heinemann, 1906

*The Snake's Pass*. Dingle, Ireland: Brandon, 1990

Stoker, William Thornley. "Memorandum on Head Injuries," ca. 1890–1896. EL3 f.S874d MS, Dracula: Notes and Outline. Rosenbach Museum and Library, Philadelphia, PA

"On a Case of Subcranial Haemorrhage Treated by Secondary Trephining." *Annals of Surgery* 7 (1888): 401–9

Stoker, William Thornley, *et al.* "A Contribution to the Surgery of the Brain." *Dublin Journal of Medical Science* 90 (1890): 292–9, 487–95

Strachey, James. "Editor's Introduction" to Josef Breuer and Sigmund Freud, *Studies on Hysteria*, trans. and ed. James Strachey. New York: Basic Books, 1957

Sully, James. "Genius and Insanity." *The Nineteenth Century* 27 (1885): 948–69

Sweet, Matthew. "Introduction" to Wilkie Collins, *The Woman in White*. New York: Penguin, 1999

Tambling, Jeremy. "*Middlemarch*, Realism, and the Birth of the Clinic." *ELH* 57.4 (1990): 939–60

Taylor, Jenny Bourne. *In the Secret Theatre of Home: Wilkie Collins, Sensation Narrative, and Nineteenth-Century Psychology*. New York: Routledge, 1988

Taylor, Jenny Bourne and Sally Shuttleworth, eds. *Embodied Selves: An Anthology of Psychological Texts, 1830–1890*. New York: Oxford University Press, 1998

Thomas, Ronald. *Detective Fiction and the Rise of Forensic Science*. Cambridge University Press, 1999

Turner, Mark. "The Cognitive Study of Art, Language, and Literature." *Poetics Today* 23.1 (2002): 9–20

Tyler, Thomas. "Review of *The Man of Genius*, by Cesare Lombroso." *The Academy* (March 20, 1892): 303

Vasbinder, Samuel Holmes. *Scientific Attitudes in Mary Shelley's* Frankenstein. Ann Arbor: University of Michigan Research Press, 1984

Veeder, William. "Children of the Night: Stevenson and Patriarchy." In William Veeder and Gordon Hirsch, eds., *Dr. Jekyll and Mr. Hyde after One Hundred Years*. University of Chicago Press, 1988. 107–60

Vyver, Bertha. *Memoirs of Marie Corelli*. London: Alston Rivers Ltd., 1930

Wagar, W. Warren. *H. G. Wells: Traversing Time*. Middletown, CT: Wesleyan University Press, 2004

Walters, J. Cuming. "Epilogue" to Bertha Vyver, *Memoirs of Marie Corelli*. London: Alston Rivers Ltd., 1930

Wells, H. G. *Anticipations of the Reaction of Mechanical and Scientific Progress upon Human Life and Thought*. New York: Harper & Brothers, 1902

"Appendix I: *The Chronic Argonauts*." In *The Time Machine: An Invention. A Critical Text of the 1895 London First Edition, With an Introduction and Appendices*, ed. Leon Stover. Jefferson, NC: McFarland, 1996. 174–95

*Certain Personal Matters*. London: T. Fisher Unwin, 1901

*The Correspondence of H. G. Wells*, ed. David Smith. London: Pickering & Chatto, 1998

*Experiment in Autobiography: Discoveries and Conclusions of a Very Ordinary Brain (Since 1866)*. 2 vols. Boston: Faber and Faber, 1984

*The First Men in the Moon*, ed. Patrick Parrinder. New York: Penguin, 2005

*H. G. Wells: Early Writings in Science and Science Fiction*, ed. Robert Philmus. Berkeley: University of California Press, 1975

*H. G. Wells: Journalism and Prophecy 1893–1946*, ed. W. Warren Wagar. Boston: Houghton Mifflin, 1964

"Introduction" to *The Scientific Romances of H. G. Wells*. London: Victor Gollancz, 1933

*The Invisible Man*, ed. Patrick Parrinder. New York: Penguin, 2005

*The Island of Dr. Moreau*, ed. Patrick Parrinder. New York: Penguin, 2005

*The Island of Dr. Moreau: A Variorum Text*, ed. Robert Philmus. Athens: University of Georgia Press, 1993

"Review of *The Human Machine*, by John Ferguson Nisbet." *The Academy* (May 6, 1899): 502

"Scepticism of the Instrument." *Mind* 13.51 (July 1904): 379–93

"Science, in School and after School." *Nature* 50 (September 1894): 525–6

"Science Teaching: An Ideal and Some Realities." *Educational Times* 48 (January 1895): 23–9

*The War of the Worlds*, ed. Patrick Parrinder. New York: Penguin, 2005

Wells, H. G., Julian S. Huxley, and G. P. Wells. *The Science of Life*. New York: Doubleday, 1934

Wicke, Jennifer. "Vampiric Typewriting: Dracula and Its Media." *ELH* 59.2 (1992): 467–93

Wilde, Oscar. "The Decay of Lying: An Observation." In *The Soul of Man under Socialism and Selected Critical Prose*, ed. Linda Dowling. New York: Penguin, 2001

Willis, Chris. "The Detective's *Doppelgänger*: Conflicting States of Female Consciousness in Grant Allen's Detective Fiction." In William Greenslade and Terence Rodgers, eds., *Grant Allen: Literature and Cultural Politics at the* Fin de Siècle. Burlington, VT: Ashgate, 2005. 143–53

Wozniak, Heather. "Brilliant Gloom: The Contradictions of British Gothic Drama, 1768–1823." Ph.D. dissertation, University of California, Los Angeles, 2008

Young, Robert M. *Mind, Brain and Adaptation in the Nineteenth-Century: Cerebral Localization and Its Biological Context from Gall to Ferrier.* New York: Oxford University Press, 1990

Zola, Émile. "The Experimental Novel." In *The Experimental Novel and Other Essays*, trans. Belle Sherman. New York: Cassell Publishing, 1893

# Index


absinthe, 159
albinism, 138, 142
alchemy, 70, 165
Aldini, Giovanni, 171
aliens (*see also* extraterrestrials), 20, 122–24, 135, 142, 145, 148, 154, 155
Allen, Grant, 3, 22, 23, 41, 85–112, 114, 115, 157, 208–13
*The Colour Sense*, 87, 94, 96, 105, 210, 213
*The Evolution of the Idea of God*, 89
*Hilda Wade*, 86, 100, 101, 208, 211, 212
*Miss Cayley's Adventures*, 86, 208
"A Mysterious Occurrence in Piccadilly," 94
"Our Scientific Observations of a Ghost," 94
*Physiological Aesthetics*, 85, 87, 91, 96, 105, 110, 212
*Recalled to Life*, 23, 86–90, 92–96, 99–101, 103, 104, 106, 107, 110, 111, 114, 115, 208
"Sight and Smell in Vertebrates," 97, 101
*What's Bred in the Bone*, 86, 208
*The Woman Who Did*, 86
ambidextrous education, 35, 39, 131
American Psychiatric Association, 188
amnesia, 90, 95, 100, 104, 111, 114
Anderson, Hans Christian, 137
anesthesia, 2, 12, 67, 68, 70, 72, 73
Anna O., 31
anthropology, 5, 8, 10, 38, 89
Anti-vivisection Act, 12, 61, 67, 98, 140
antivivisection movement, 68, 72
aphasia, 2, 90, 100, 111, 113, 211
asylums (*see also* madhouses), 47
atheism, 53, 89, 160, 165, 183
Atlantic Cable, 170
atom bomb, 157, 223
attention deficit disorder, 190
*Austin Powers*, 154
automata, 51, 52, 54, 55, 69, 72, 75, 86, 186
automatism (*see also* reflex action of the cerebrum, unconscious cerebration), 6, 51–56, 60, 78, 85, 86, 92, 95, 185, 186, 201, 207
autopsies, 2, 35
Azam, Eugène, 28, 41

Bain, Alexander, 4, 86, 110, 131
Balfour, Arthur, 74
Ball, Alan, 81
Barrett, William, 70, 167
Bazin, André, 106
"The Ontology of the Photographic Image," 106
Beard, George, 175
Beckford, William Thomas, 45
Bell, Alexander Graham, 93
Benjamin, Walter, 108
Benson, Herbert, 229
*The Relaxation Response*, 229
Bentley, George, 21, 166, 225
Bernard, Claude, 4, 55
Bicêtre, 128
Binet, Alfred, 100, 122
biological determinism, 1, 9, 22, 32, 51, 55, 81, 115, 128, 184, 188, 190, 207
bipolar disorder, 190
Blake, William, 166
blind chess players, 100
Böll, Franz, 94, 95, 105, 209
Bouillard, Jean Baptiste, 121
Bourru, Hippolyte, 31
Braddon, Mary Elizabeth, 104, 212
brain, 1, 2, 4–6, 9–17, 19–24, 29–40, 43–45, 47, 49, 51–57, 61–65, 67–70, 75, 77, 78, 80, 86, 88, 89, 92–98, 102, 103, 105, 111–13, 115, 119, 121–24, 128, 131–35, 141, 142, 143, 145–47, 149, 150, 152, 153, 156–60, 167–70, 174, 175, 177, 178, 181, 184–86, 188–90, 193, 198, 199, 203, 214, 215, 224, 230

brain cells, 1, 158, 159, 175, 178, 224
brain hemisphere asymmetry, 5, 19, 22, 29, 31, 131
brain surgery, 56, 77
brain tumors, 2, 61, 67
brain-computer interfacing (BCI), 187
*Brain: A Journal of Neurology*, 15, 45, 52
Breuer, Josef, 15, 31, 111
*Studies in Hysteria*, 111
Bridgman, Laura, 100
British Association for the Advancement of Science, 28
*British Medical Journal*, 61
Broca, Paul, 2, 35, 121, 122, 199
Brontë, Charlotte
*Jane Eyre*, 12
*Villette*, 12
Brown-Séquard, Charles Édouard, 5, 33, 35, 38, 40, 131, 150
Brown, Charles Brockden, 11
Bruce, Lewis, 6, 29, 192
Buddhism, 162, 225
Burdon Sanderson, John Scott, 1, 55, 56, 64, 73, 77
*Handbook of the Physiological Laboratory*, 73
Burot, P., 31
Burton, Tim, 154
*Mars Attacks*, 154
Bush, George, 189
Byron, George Gordon, 126

Caine, Hall, 76, 157
Cajal, Santigo Ramón y, 9, 20, 159
Carpenter, William Benjamin, 5, 11, 54–56, 77, 86, 174, 203, 207, 215, 228
*Principles of Human Physiology*, 54
*Principles of Mental Physiology*, 174, 203
case studies, 16, 19, 28–30, 43–48, 59, 62
catalepsy, 59, 60
cerebral cortex, 2, 14, 21, 67, 97, 102, 158
cerebral hemispheres, 24, 34, 51, 61, 174
cerebral localization, 2, 3, 5, 6, 10–13, 19, 21–23, 34, 35, 48, 51, 52, 54–56, 61, 65, 67, 80, 97, 98, 115, 121, 122, 124, 148, 156, 158, 159, 174, 176, 177, 184–86, 188, 193, 203, 216
Charcot, Jean Martin, 5, 28, 55, 56, 67, 68, 71, 72, 73, 74, 77
Cheyne, George, 59
*The English Malady*, 59
*The Children of the Damned*, 186
Christian Science, 163, 165, 166
Christianity, 24, 57, 89, 161, 165
Clemens, Samuel, 76
Clifford, William Kingdon, 52–54, 86, 115
Cobbe, Frances Power, 12, 67, 68, 73, 206
"Science in Excelsis," 72, 73
*The Modern Rack: Papers on Vivisection*, 12, 67
cognitive literary criticism, 4, 189, 190, 230
Collins, Wilkie, 13, 69, 104, 127, 218
*Heart and Science*, 13, 69, 127, 218
*The Moonstone*, 104, 212
Combe, George, 11
Corelli City, Colorado, 164
Corelli, Marie, 3, 17, 20–24, 82, 156–84
*Ardath*, 158, 164, 165, 171, 177
Heliobas Trilogy, 158
*The Life Everlasting*, 20, 158, 162, 163, 165, 166, 169, 170, 180, 182, 183
*The Mighty Atom*, 165
*A Romance of Two Worlds*, 8, 17, 20, 21, 24, 158, 160, 161, 169, 176, 223, 225
*The Secret Power*, 180, 223
*The Sorrows of Satan*, 159
*The Soul of Lilith*, 158, 159, 177
*The Treasure of Heaven: A Romance of Riches*, 157
*Wormwood*, 158, 159
*The Young Diana*, 156, 165, 178, 180, 181
*Cornhill Magazine*, 29, 40, 43, 211
corpus callosotomy, 37
cortical maps, 2, 12, 61, 62, 64
craniometry, 36, 122, 123, 124
Crichton, Michael, 185, 186, 229
*Terminal Man*, 185, 186, 229
criminology, 8, 56, 86, 202
Cronenberg, David, 187
*eXistenZ*, 187
Crookes, William, 169, 227
Curie, Marie, 180, 181
"Radium and Radioactivity," 181
Curie, Pierre, 180
Cuvier, Georges, 127
cyberpunk, 186, 188, 230

Dallas, E. S., 4
Dames, Nicholas, 4, 104
Darwin, Charles, 10, 12, 20, 38, 51, 66, 88, 89, 96, 97, 99, 105, 121, 130, 131, 190
*The Descent of Man*, 13, 38, 66
*The Origin of Species*, 13, 88
Daudet, Léon, 71
depression, 111, 154, 161, 175, 189, 190
Descartes, René, 2, 34, 51, 52, 75, 85, 86, 207, 208
detective fiction, 86, 87, 92, 101, 104, 106, 108, 212
*Diagnostic and Statistical Manual of Mental Disorders* (DSM), 31, 188, 189, 230

Dickens, Charles, 111
*Dictionary of National Biography*, 41
dissociation, 31, 92, 209
dissociative identity disorder (*see* multiple personality), 31, 209
double brain, 30, 33, 40, 48, 197
Doyle, Arthur Conan, 76, 86, 87, 89, 91, 99, 107, 208, 209, 211
  *The Crooked Man*, 104
  *The Dancing Men*, 104
  "A Scandal in Bohemia," 87, 104
  *A Study in Scarlet*, 87
Du Bois-Reymond, Emil, 174
dual consciousness (*see also* dual personality), 32, 35, 100, 211
dual personality (*see also* dual consciousness), 5, 27–31, 33, 34, 37, 38, 40–42, 44, 47, 49, 90, 92, 100, 209
dynamo, 173
dystopia, 20, 23, 119, 123, 132, 144, 153, 184, 188

Eastman, George, 107
Eddy, Mary Baker, 166
Edison, Thomas, 173, 174
electricity, 110, 156, 158–62, 167, 168, 170–83, 214, 223, 227
  electric battery, 171, 174–78
  electric circuit, 171, 176
  Electric Creed, 158, 159, 161–65, 171, 180, 181, 183, 223
  electric fish, 172, 173, 176, 177, 227
  electric shock, 172, 176
  electrotherapy, 161, 175, 178, 182
Eliot, George, 3, 4, 12, 76, 132, 191, 194
  *Daniel Deronda*, 4
  *The Lifted Veil*, 12
  *Middlemarch*, 3, 8
  *The Mill on the Floss*, 8
Ellis, Havelock, 15
epilepsy, 128, 189
epiphenomenalism, 52, 201
Esquirol, Jean, 128
eugenics, 101, 134
evolution, 5, 7, 8, 20, 37, 38, 77, 88, 96, 101, 119, 123, 131, 132, 136, 141, 145–47, 151–53, 188, 222
evolutionary regression, 56, 128, 146
extraterrestrials (*see also* aliens), 20, 119, 120, 130, 133, 143, 148, 154, 170

Faust legends, 17, 70
Félida X, 28, 31, 35, 40, 41, 42, 43, 44, 100, 197, 211
Ferrier, David, 1, 2, 5, 12–16, 21, 35, 38, 52, 55, 56, 61, 62, 64–71, 73, 77, 97, 98, 124, 125, 189, 191, 193, 203, 204, 205, 207
  "The Brain of the Criminal Lunatic," 15, 38
  *The Functions of the Brain*, 64, 69, 204
  *The Localization of Cerebral Disease*, 66
Ferrier trial, 13, 21, 23, 24, 61, 67–69, 98
First World War, 80, 157
fissure of Rolando, 62
Fliess, Wilhelm, 32
Flourens, Jean Pierre Marie, 10
Forster, Thomas, 172
Foucault, Michel, 46, 126, 194
Fowler, O. S., 11
freak shows, 142
free will, 1, 2, 11, 50, 51, 54, 69, 79, 185
Freud, Sigmund, 15, 31, 32, 48, 74, 111, 115, 167, 179
  *Project for a Scientific Psychology*, 179, 198
  *Studies in Hysteria*, 111
Fritsch, Gustav, 2, 12, 52, 62, 69, 124, 158, 174
  "On the Electrical Excitability of the Cerebrum," 174
frontal lobes, 53, 121, 122, 125, 130, 138, 150

Gall, Franz Joseph, 11, 30, 34, 35, 49, 51, 121, 193
Galton, Francis, 100, 108, 120, 122, 126–28, 130, 133, 134, 135, 138, 154
  *English Men of Science: Their Nature and Nurture*, 127, 134
  *Hereditary Genius*, 126, 130
Galvani, Luigi, 171–74, 227
  *De viribus electricitatis in motu musculari. Commentarius*, 171
galvanism, 176
genius, 125, 127, 129, 134
germ plasm theory of inheritance, 132
Gibson, William, 186
  *Neuromancer*, 186
Gilman, Charlotte Perkins, 106, 177
  "The Yellow Wallpaper," 106, 177
Gladstone, William, 157
Goethe, Johann Wolfgang von, 172
Golgi, Camillo, 159
Goltz, Frierich, 12, 67, 205
Gothic fiction, 7, 9, 11, 14, 16, 30, 69, 186, 192, 193
Gothic romance, 1, 3, 5, 8–10, 16, 19, 20, 22, 24, 30, 47, 154, 160, 193, 223
Gould, Steven J., 35
Grand Lunar, 20, 152, 153
Gray, Robert, 59
  *The Theory of Dreams*, 59
Gurney, Edmund, 167, 226
  *Phantasms of the Living*, 167, 169

Hacking, Ian, 28, 41, 48
Haggard, H. Rider, 7, 18
  *King Solomon's Mines*, 17

Hall, G. Stanley, 74, 167
handedness (left vs. right), 40, 131, 199
handwriting, 40
Haraway, Donna, 230
"Cyborg Manifesto," 187
Harrington, Anne, 32, 34, 36, 38, 197
Harris, Charlaine, 81
Hawthorne, Nathaniel, 7
Helmholtz, Hermann von, 105, 106, 173, 174, 175, 212, 213
*Physiological Optics*, 105
hemorrhage, 27, 61, 62
Henley, William Ernest, 27
Hertz, Heinrich, 168
hibernation, 60
Hitzig, Eduard, 2, 12, 52, 62, 69, 124, 158, 174, 193
"On the Electrical Excitability of the Cerebrum," 174
Hodgson, Shadworth, 52, 86
Hogg, James, 45
*Confessions of a Justified Sinner*, 45
Holland, Henry, 5, 30, 33, 35
Holmes, Oliver Wendell, 54, 86, 214
Holmes, Sherlock, 86, 87, 90, 91, 99, 104, 106
Hope, Anthony, 18
Hopkins, Gerard Manley, 42
Howells, William Dean, 18
Hurley, Kelly, 5, 193
Huxley, Julian, 122
*The Science of Life*, 122, 138, 146, 149, 150, 222
Huxley, Thomas Henry, 20, 28, 52–54, 56, 78, 86, 89, 96, 115, 119, 120, 122, 123, 126, 134, 201
"On the Hypothesis that Animals are Automata, and its History," 28, 52
Huysmans, Joris Karl, 130
hypermnesia, 111
hypnotism, 59, 71, 74
hysteria, 15, 37, 44, 71, 128, 175, 200, 214

IQ tests, 122, 126
Ibsen, Henrik, 130
Industrial Revolution, 127
insects, 150, 222
insomnia, 161, 175
International Medical Conference 1881, 67
*Invasion of the Body Snatchers*, 186

Jack the Ripper, 94
Jackson, John Hughlings, 2, 12, 14, 35, 68, 193
James, Henry, 13, 17, 18, 21
"The Art of Fiction," 18
James, William, 13, 14, 51, 52, 74, 77, 167, 201
"Frederic Myers's Service to Psychology," 13
Janet, Pierre, 30, 31, 74, 167, 197
*Psychological Automatism*, 31
Jenkin, Fleeming, 30

Kent, Constance, 100
Kinetoscope, 110
Kingsford, Anna, 71
Kleinman, Arthur, 190
*Knowledge: An Illustrated Magazine of Science*, 41
Krafft-Ebing, Richard von, 10, 15
Kuhn, Thomas, 64
Kühne, F. Wilhelm, 94

Lamarck, Jean Baptiste, 121
*Zoological Philosophy*, 20, 121
Lamarck's first law, 20, 121, 132, 150
Lamarck's second law, 121
Lamarckism, 123, 131, 132, 216
*Lancet, The*, 27
Lang, Andrew, 7, 74, 160
Lankester, Edwin Ray, 146
*Degeneration: A Chapter in Darwinism*, 146
Laycock, Thomas, 54
Le Bon, Gustav, 132
Le Gallienne, Richard, 99
Lee, Sarah, 60
*Anecdotes of the Habits and Instincts of Animals*, 60
Lewes, George Henry, 4
Leyden jar, 179, 228
Lieber, Charles, 187
Liepmann, Hugo, 35
literary Darwinism, 190
Lodge, Oliver, 70, 75, 76, 167, 168, 169
Lombroso, Cesare, 74, 120, 127–31, 133, 134, 139, 154, 167
*The Man of Genius*, 127, 128
*The Los Angeles Times*, 188
Luddites, 46, 57, 73
Luria, A. R., 14, 15, 16
*The Mind of a Mnemonist*, 15

Mackay, Charles, 164, 226
*Memoirs of Extraordinary Popular Delusions and the Madness of Crowds*, 164
mad scientist, 55, 70, 119–23, 127, 130, 133, 136, 137, 141, 143, 144, 148, 154, 155, 218
madhouses, 1, 14, 62, 64, 77
magnetic resonance imaging (MRI), 189, 190
Marey, Étienne Jules, 112
Marsh, Richard, 104, 212
Martians, 102, 144, 145, 146
Marvin the Martian, 155
*The Matrix*, 178, 228
Maturin, Charles, 45
*Melmoth the Wanderer*, 45

Maudsley, Henry, 33, 38, 97, 99, 100, 102, 132, 139
*Physiology and Pathology of Mind*, 99
"The Double Brain," 33
Meissner, Paul Traugott, 174
memory
eidetic memory (*see* photographic memory), 99
photographic memory, 95, 99, 100, 103, 113, 211
visual memory, 15, 23, 93, 98–100, 106
mesmerism, 11, 164, 192, 225
Mesnet, Ernest, 28
Meyer, Stephenie, 81
microelectrode, 179
microscope, 87, 99, 101, 144, 211
electron microscope, 179
Mighall, Robert, 5, 30, 56, 193
migraine, 175
*Mind*, 15, 45, 52, 86, 96, 97, 124
mind control, 186, 187, 188
mind–body divide, 1, 51
Mitchell, Silas Weir, 177
Moleschott, Jacob, 173
monomania, 128, 139, 141
moral insanity, 100, 129, 139, 141, 144, 147
moral managers, 38, 52
Moreau, Jacques, 100, 120, 128, 134
*Morbid Psychology*, 128, 129, 134, 219
multiple personality, 6, 31, 34, 49, 198, 209
Muybridge, Eadweard, 23, 108, 110–13, 213, 214, 215
*Animal Locomotion*, 108, 112
Myers, A. T., 31
"The Life-History of a Case of Double or Multiple Personality," 31
Myers, Frederic W. H., 6, 14, 27, 28, 29, 30, 33, 40, 43, 70, 75, 76, 78, 167, 196, 197, 198
*Human Personality and its Survival of Bodily Death*, 75
"Multiplex Personality," 28, 29, 31, 33, 43, 192, 198

Nancy Drew novels, 104
nanotechnology, 187, 189
natural selection, 20, 85, 88, 119, 121, 131, 145
neurasthenia, 20, 161, 175, 176, 178, 183, 225, 228
neuron, 9, 20, 24, 156, 158, 159, 175, 176, 178–80, 183, 184, 189, 224
axon, 158, 159
dendrite, 159
neuron doctrine, 9, 20, 24, 156, 158, 159, 178, 179, 183, 224
neuroses, 175
New Thought, 166
New Woman, 86, 104, 161
Newton, Isaac, 127
Nisbet, John Ferguson, 100, 124, 125, 126, 128, 133, 134, 141, 154
*The Human Machine*, 133
*The Insanity of Genius*, 124, 125, 134
Nobel Prize, 159
Nordau, Max, 128, 130, 133, 134, 141
*Degeneration*, 128, 134
normal distribution, 126
Normal School of Science, South Kensington, 20, 119, 138

objectivity, 3, 10, 15, 16, 19, 21, 30, 45, 46, 48, 79, 92, 106, 108, 164, 167, 194
ophthalmology, 87, 94, 208
optic nerve, 23, 88, 89, 92

Paley, William, 88, 89, 105
*Natural Theology*, 88
parallelism, 53, 54, 69
parasitism, 146
Paris Electricity Exhibition of 1881, 173
Pascal, Blaise, 127
pelvis, 122, 123, 149
Phelps, Elizabeth Stuart, 163
*The Gates Ajar*, 163
phonograph, 57, 93, 96, 210
photography, 90, 92, 93, 95, 96, 106, 107, 108, 110, 111, 114, 213
camera, 23, 87–90, 92, 93, 95, 96, 103, 105, 107, 108, 110, 112, 115
chronophotography, 23, 107, 108, 110, 112, 113, 213
composition photography, 96, 210
flash photography, 93, 105, 107, 110–12, 214
negatives, 88, 96
photographic gun, 112
phrenology, 11, 12, 21, 34, 65, 121, 192–94
physiologists, 1, 5, 30, 53, 56, 71–73, 77, 79, 93, 94, 106, 150, 170, 206
physiology, 1, 52, 55, 64, 79, 86, 87, 93, 96, 115, 123, 174, 190, 202
pineal gland, 2, 34, 51, 208
Pinel, Philippe, 128
Planché, James Robinson, 60
*The Vampire*, 60
Podmore, Frank, 167
Poe, Edgar Allan, 7, 11
"The Fall of the House of Usher," 12
"The Imp of the Perverse," 11
"The Murders of Rue Morgue," 12
Polidori, John, 60, 61
*The Vampyre*, 60
post-traumatic stress (*see* traumatic shock), 111

Prince of Wales, 157
Proctor, Richard, 28, 29, 35, 37, 40, 41, 43, 44, 89, 200, 211
"Dual Consciousness," 29, 41
"Have We Two Brains?" 29, 41
proprioception, 14, 36
psychical research, 6, 13, 14, 70, 74–80, 94, 168–71, 192
psychoanalysis, 48
*Punch*, 75, 76, 119, 143, 145, 155

Queen Victoria, 157
Quételet, Adolphe, 126
*A Treatise on Man and the Development of his Faculties*, 126

Radcliffe, Ann, 7, 8
*The Mysteries of Udolpho*, 8
radioactivity, 24, 178, 180, 182–84
radium, 162, 180–82, 184, 229
railway spine, 111, 214
realism, 3, 4, 13, 21, 64, 81, 191, 212
reflex, 6, 10, 22, 52, 54, 56, 65, 95
reflex action of the cerebrum (*see also* automatism, unconscious cerebration), 54, 95
reincarnation, 164, 171, 225
rest cure, 177, 180, 184
retina, 92–96, 106
Reynolds, Mary, 41, 199
Ribot, Théodule, 92, 95, 97, 99, 100–12, 114, 209, 211, 215
*Diseases of Memory*, 92, 95, 111, 211
Richardson, Alan, 11, 189
Rockwell, A. D., 175
*The Rocky Horror Picture Show*, 154
Roentgen Society of London, 181
romance, 1, 3–9, 13, 16–21, 23, 24, 45, 46, 123, 134, 148, 149, 156, 157, 160, 161, 162, 163, 164, 168, 170, 171, 176, 180, 181, 184, 192, 195
Rosicrucianism, 163, 165
Royal College of Surgeons in Ireland, 6, 58

Sacks, Oliver, 14, 15, 16, 36, 185, 194
*Salpêtrière*, 67, 128
Sand, George, 132
schizophrenia, 154, 223
science fiction, 17, 19, 23, 55, 124, 154, 156, 157, 185, 186, 187, 188, 223
scientific materialism, 75, 153, 156, 165, 184
scientific method, 12, 20, 35, 46, 76, 78, 123, 193
scientific romance, 17, 19, 20, 23, 119, 122, 123, 133–35, 139, 142, 144, 148, 149, 195
Scotland Yard, 93, 94, 102, 106
Scott, Walter, 7
*Ivanhoe*, 7
*Waverley*, 7
Second World War, 179
Selenites, 143, 148, 149, 150, 151, 152, 153
senile dementia, 114
sensation novels, 5, 18, 92, 104
Sergeant F., 28, 29, 40–44, 47, 53, 78
sexology, 5, 8, 10, 56, 194, 202
Shelley, Mary, 17, 45, 55, 121, 126, 172
*Frankenstein*, 17, 45, 54, 55, 121, 126, 127, 139, 148, 172, 174, 185, 227
Shelley, Percy, 126
*Prometheus Unbound*, 126
Sherrington, Charles, 65, 179
shilling-shocker, 19
Sidgwick, Henry, 74
Siemens, Werner von, 173
skull, 2, 11, 35, 36, 63, 67, 122, 123, 142, 146, 149, 150, 152
sleep, 19, 36, 59, 60
sleepwalking (*see* somnambulism), 6, 50, 59, 60
Society for Psychical Research, 74, 75, 77, 78, 167, 169, 170
*Proceedings of the Society for Psychical Research*, 167
*The Journal of the Society for Psychical Research*, 60, 74, 78
somnambulism, 59, 60, 78
soul, 2, 9–12, 30, 34, 37, 47, 49–53, 55–57, 60, 61, 69, 75, 77, 79–81, 85, 86, 89, 94, 115, 161–64, 167, 169, 171, 172, 177, 178, 180–82, 184, 185, 209
Spielberg, Steven, 154
*The War of the Worlds* (2005), 154
Spencer, Herbert, 4, 85, 96, 97, 101, 102, 110, 114, 134, 210
spinal cord, 172
spiritualism, 23, 76, 80, 163, 164, 183, 209, 225
séances, 74, 164
Spurzheim, Johann Gaspar, 11
St. Patrick's Hospital, Dublin, 61
Stevenson, Fanny, 27
Stevenson, Robert Louis, 3, 5–7, 13, 14, 17–24, 27–35, 37–49, 121, 130, 156, 160, 161, 184
*Deacon Brodie, or the Double Life*, 27
"A Gossip on Romance," 19
"A Humble Remonstrance," 18
"Markheim," 27, 44
*Strange Case of Dr. Jekyll and Mr. Hyde*, 3, 7, 8, 16, 17, 19, 22, 27–34, 36, 37, 40, 44, 45, 48, 49, 76, 121, 127, 185, 190–93
*Treasure Island*, 17

Stoker, Bram, 1, 3, 5, 6, 13, 14, 22, 23, 50, 55–62, 64, 69, 71, 73, 75–78, 85, 86, 156, 165, 192, 203
*Dracula*, 1, 3, 5–8, 10, 17, 22, 50, 51, 53, 55–57, 59–62, 64, 69–73, 76–81, 146, 185, 186, 190, 202
*Personal Reminiscences of Henry Irving*, 71, 76
*The Snake's Pass*, 58
Stoker, George, 58
Stoker, Thornley, 6, 58, 61, 62, 64, 65
"On a Case of Subcranial Haemorrhage Treated by Secondary Trephining," 62
*Strand Magazine*, 87
Sully, James, 30, 33, 197
"The Dream as Revelation," 33
"Genius and Insanity," 127
Swedenborgians, 166
Swift, Jonathan, 61, 151
*Sybil*, 31
sympathetic vibration, 168, 169
synapse, 179, 189

Talbot, William Henry Fox, 107
Taylor, Jenny Bourne, 5
telegraphy, 57, 157, 158, 168, 223
telekinesis, 177
telepathy, 14, 64, 70, 73, 74, 78, 163, 167–70, 177
telephone, 93, 168, 170, 224
Terman, Lewis, 130
Thackeray, William Makepeace, 4
*Vanity Fair*, 4
theosophy, 162
thermodynamics, 158, 171, 173, 183, 223
conservation of energy, 173, 175, 179, 181, 183
*The Three Faces of Eve*, 31
trance, 6, 59, 60, 75, 78, 209
traumatic shock, 91, 92, 103, 111, 113
trepanning, 61, 62
Trinity College, Dublin, 5, 57, 58, 192
*True Blood*, 81
*Twilight*, 81

unconscious cerebration (*see also* automatism, reflex action of the cerebrum), 6, 54, 56, 95
United States Department of Energy, 155
utopia, 180

vampires, 6, 10, 22, 50, 51, 55–60, 64, 69, 70, 72, 73, 76–82, 85, 185, 186
vampire trap, 60
Verne, Jules, 157
Viagra, 81
Victoria Street Society, 12, 67
Villiers de L'Isle-Adam, Auguste, 173
*Tomorrow's Eve*, 174
vision, 86–89, 92, 96–98, 102, 103, 105, 106, 110, 112, 115, 119, 208, 213
blind spots, 105
floaters, 105, 212
retinal after-images, 93, 94, 105, 110, 213
Vivet, Louis, 27–29, 31, 33, 43, 49, 198
vivisection, 1, 61, 64, 67, 68, 72, 73, 77, 78, 85, 98, 139–41, 211
Volta, Alessandro, 171, 172, 174

Wagner, Richard, 130
Walpole, Horace, 7, 8, 9
*Castle of Otranto*, 7
Walters, John Cuming, 163
Weismann, August, 131
Weismannism, 132
Wells, George Philip, 122
*The Science of Life*, 122, 138, 146, 149, 150, 222
Wells, H. G., 3, 13, 14, 17, 19–24, 69, 102, 119, 120, 122–24, 127, 128, 130–45, 147–57, 160, 161, 184
*Anticipations*, 134, 140, 144, 148, 149, 221
The Chronic Argonauts, 136, 138
"The Empire of the Ants," 150
*Experiment in Autobiography*, 137, 153, 220
*The First Men in the Moon*, 19, 20, 23, 119, 132, 133, 135, 137, 143, 144, 148, 149, 221
*The Food of the Gods*, 17
"Human Evolution, an Artificial Process," 132
"Incidental Thoughts on a Bald Head," 132
*The Invisible Man*, 119, 123, 133, 136, 138, 141, 142
*The Island of Doctor Moreau*, 8, 13, 17, 69, 119, 134, 138, 139, 141, 148
"The Man of the Year Million," 20, 119, 122, 135, 137, 143, 145, 146, 148, 149, 151
*A Modern Utopia*, 134, 219
*The Science of Life*, 119, 121, 125, 146, 150, 222
*The Time Machine*, 19, 123, 134, 136, 152, 219, 221
*The War of the Worlds*, 3, 20, 23, 102, 119, 132, 133, 135, 137, 143, 145–50, 190
*World Brain*, 153
West Riding Lunatic Asylum, 65
Wigan, Arthur Ladbroke, 5, 30, 33, 35, 38, 40
*A New View of Insanity: The Duality of the Mind*, 35
Wilde, Oscar, 27, 58, 141, 157, 224
Wilde, William, 58
*Irish Popular Superstitions*, 58

willpower, 22, 53, 69, 174, 182
Wilson, Edward O., 190
Wolpaw, Jonathan, 188
Women's Suffrage, 161

X-rays, 157, 223

Yeo, Gerald Francis, 68
*Young Frankenstein*, 154

Zola, Émile, 3
zombies, 81, 207
zoopraxiscope, 108, 113, 213

CAMBRIDGE STUDIES IN
NINETEENTH-CENTURY LITERATURE AND CULTURE

General editor
GILLIAN BEER, *University of Cambridge*

Titles published

1. The Sickroom in Victorian Fiction: The Art of Being Ill
MIRIAM BAILIN, *Washington University*

2. Muscular Christianity: Embodying the Victorian Age
edited by DONALD E. HALL, *California State University, Northridge*

3. Victorian Masculinities: Manhood and Masculine Poetics in Early Victorian Literature and Art
HERBERT SUSSMAN, *Northeastern University, Boston*

4. Byron and the Victorians
ANDREW ELFENBEIN, *University of Minnesota*

5. Literature in the Marketplace: Nineteenth-Century British Publishing and the Circulation of Books
edited by JOHN O. JORDAN, *University of California, Santa Cruz* and ROBERT L. PATTEN, *Rice University, Houston*

6. Victorian Photography, Painting and Poetry
LINDSAY SMITH, *University of Sussex*

7. Charlotte Brontë and Victorian Psychology
SALLY SHUTTLEWORTH, *University of Sheffield*

8. The Gothic Body: Sexuality, Materialism and Degeneration at the *Fin de Siècle*
KELLY HURLEY, *University of Colorado at Boulder*

9. Rereading Walter Pater
WILLIAM F. SHUTER, *Eastern Michigan University*

10. Remaking Queen Victoria
edited by MARGARET HOMANS, *Yale University* and ADRIENNE MUNICH, *State University of New York, Stony Brook*

11. Disease, Desire, and the Body in Victorian Women's Popular Novels
PAMELA K. GILBERT, *University of Florida*

12. Realism, Representation, and the Arts in Nineteenth-Century Literature
ALISON BYERLY, *Middlebury College, Vermont*

13. Literary Culture and the Pacific
VANESSA SMITH, *University of Sydney*

14. Professional Domesticity in the Victorian Novel: Women, Work and Home
MONICA F. COHEN

15. Victorian Renovations of the Novel: Narrative Annexes and the Boundaries of Representation
SUZANNE KEEN, *Washington and Lee University, Virginia*

16. Actresses on the Victorian Stage: Feminine Performance and the Galatea Myth
GAIL MARSHALL, *University of Leeds*

17. Death and the Mother from Dickens to Freud: Victorian Fiction and the Anxiety of Origin
CAROLYN DEVER, *Vanderbilt University, Tennessee*

18. Ancestry and Narrative in Nineteenth-Century British Literature: Blood Relations from Edgeworth to Hardy
SOPHIE GILMARTIN, *Royal Holloway, University of London*

19. Dickens, Novel Reading, and the Victorian Popular Theatre
DEBORAH VLOCK

20. After Dickens: Reading, Adaptation and Performance
JOHN GLAVIN, *Georgetown University, Washington, DC*

21. Victorian Women Writers and the Woman Question
edited by NICOLA DIANE THOMPSON, *Kingston University, London*

22. Rhythm and Will in Victorian Poetry
MATTHEW CAMPBELL, *University of Sheffield*

23. Gender, Race, and the Writing of Empire: Public Discourse and the Boer War
PAULA M. KREBS, *Wheaton College, Massachusetts*

24. Ruskin's God
MICHAEL WHEELER, *University of Southampton*

25. Dickens and the Daughter of the House
HILARY M. SCHOR, *University of Southern California*

26. Detective Fiction and the Rise of Forensic Science
RONALD R. THOMAS, *Trinity College, Hartford, Connecticut*

27. Testimony and Advocacy in Victorian Law, Literature, and Theology
JAN-MELISSA SCHRAMM, *Trinity Hall, Cambridge*

28. Victorian Writing about Risk: Imagining a Safe England in a Dangerous World
ELAINE FREEDGOOD, *University of Pennsylvania*

29. Physiognomy and the Meaning of Expression in Nineteenth-Century Culture
LUCY HARTLEY, *University of Southampton*

30. The Victorian Parlour: A Cultural Study
THAD LOGAN, *Rice University, Houston*

31. Aestheticism and Sexual Parody 1840–1940
DENNIS DENISOFF, *Ryerson University, Toronto*

32. Literature, Technology and Magical Thinking, 1880–1920
PAMELA THURSCHWELL, *University College London*

33. Fairies in Nineteenth-Century Art and Literature
NICOLA BOWN, *Birkbeck, University of London*

34. George Eliot and the British Empire
NANCY HENRY, *The State University of New York, Binghamton*

35. Women's Poetry and Religion in Victorian England: Jewish Identity and Christian Culture
CYNTHIA SCHEINBERG, *Mills College, California*

36. Victorian Literature and the Anorexic Body
ANNA KRUGOVOY SILVER, *Mercer University, Georgia*

37. Eavesdropping in the Novel from Austen to Proust
ANN GAYLIN, *Yale University*

38. Missionary Writing and Empire, 1800–1860
ANNA JOHNSTON, *University of Tasmania*

39. London and the Culture of Homosexuality, 1885–1914
MATT COOK, *Keele University*

40. Fiction, Famine, and the Rise of Economics in Victorian Britain and Ireland
GORDON BIGELOW, *Rhodes College, Tennessee*

41. Gender and the Victorian Periodical
HILARY FRASER, *Birkbeck, University of London*
JUDITH JOHNSTON and STEPHANIE GREEN, *University of Western Australia*

42. The Victorian Supernatural
edited by NICOLA BOWN, *Birkbeck College, London*
CAROLYN BURDETT, *London Metropolitan University* and
PAMELA THURSCHWELL, *University College London*

43. The Indian Mutiny and the British Imagination
GAUTAM CHAKRAVARTY, *University of Delhi*

44. The Revolution in Popular Literature: Print, Politics and the People
IAN HAYWOOD, *Roehampton University of Surrey*

45. Science in the Nineteenth-Century Periodical: Reading the Magazine of Nature
GEOFFREY CANTOR, *University of Leeds*
GOWAN DAWSON, *University of Leicester*
GRAEME GOODAY, *University of Leeds*
RICHARD NOAKES, *University of Cambridge*
SALLY SHUTTLEWORTH, *University of Sheffield* and
JONATHAN R. TOPHAM, *University of Leeds*

46. Literature and Medicine in Nineteenth-Century Britain from Mary Shelley to George Eliot
JANIS MCLARREN CALDWELL, *Wake Forest University*

47. The Child Writer from Austen to Woolf
edited by CHRISTINE ALEXANDER, *University of New South Wales*
and JULIET MCMASTER, *University of Alberta*

48. From Dickens to Dracula: Gothic, Economics, and Victorian Fiction
GAIL TURLEY HOUSTON, *University of New Mexico*

49. Voice and the Victorian Storyteller
IVAN KREILKAMP, *University of Indiana*

50. Charles Darwin and Victorian Visual Culture
JONATHAN SMITH, *University of Michigan-Dearborn*

51. Catholicism, Sexual Deviance, and Victorian Gothic Culture
PATRICK R. O'MALLEY, *Georgetown University*

52. Epic and Empire in Nineteenth-Century Britain
SIMON DENTITH, *University of Gloucestershire*

53. Victorian Honeymoons: Journeys to the Conjugal
HELENA MICHIE, *Rice University*

54. The Jewess in Nineteenth-Century British Literary Culture
NADIA VALMAN, *University of Southampton*

55. Ireland, India and Nationalism in Nineteenth-Century Literature
JULIA WRIGHT, *Dalhousie University*

56. Dickens and the Popular Radical Imagination
SALLY LEDGER, *Birkbeck, University of London*

57. Darwin, Literature and Victorian Respectability
GOWAN DAWSON, *University of Leicester*

58. 'Michael Field': Poetry, Aestheticism and the *Fin de Siècle*
MARION THAIN, *University of Birmingham*

59. Colonies, Cults and Evolution: Literature, Science and Culture in Nineteenth-Century Writing
DAVID AMIGONI, *Keele University*

60. Realism, Photography and Nineteenth-Century Fiction
DANIEL A. NOVAK, *Lousiana State University*

61. Caribbean Culture and British Fiction in the Atlantic World, 1780–1870
TIM WATSON, *University of Miami*

62. The Poetry of Chartism: Aesthetics, Politics, History
MICHAEL SANDERS, *University of Manchester*

63. Literature and Dance in Nineteenth-Century Britain: Jane Austen to the New Woman
CHERYL WILSON, *Indiana University*

64. Shakespeare and Victorian Women
GAIL MARSHALL, *Oxford Brookes University*

65. The Tragi-Comedy of Victorian Fatherhood
VALERIE SANDERS, *University of Hull*

66. Darwin and the Memory of the Human: Evolution, Savages, and South America
CANNON SCHMITT, *University of Toronto*

67. From Sketch to Novel: The Development of Victorian Fiction
AMANPAL GARCHA, *Ohio State University*

68. The Crimean War and the British Imagination
STEFANIE MARKOVITS, *Yale University*

69. Shock, Memory and the Unconscious in Victorian Fiction
JILL L. MATUS, *University of Toronto*

70. Sensation and Modernity in the 1860s
NICHOLAS DALY, *University College Dublin*

71. Ghost-Seers, Detectives, and Spiritualists: Theories of Vision in Victorian Literature and Science
SRDJAN SMAJIĆ, *Furman University*

72. Satire in an Age of Realism
AARON MATZ, *Scripps College, California*

73. Thinking about other People in Nineteenth-Century British Writing
ADELA PINCH, *University of Michigan*

74. Tuberculosis and the Victorian Literary Imagination
KATHERINE BYRNE, *University of Ulster, Coleraine*

75. Urban Realism and the Cosmopolitan Imagination: Visible City, Invisible World
TANYA AGATHOCLEOUS, *Hunter College, City University of New York*

76. Women, Literature, and the English Domesticated Landscape: England's Disciples of Flora, 1780–1870
JUDITH W. PAGE, *University of Florida*
ELISE L. SMITH, *Millsaps College, Mississippi*

77. Time and the Moment in Victorian Literature and Society
SUE ZEMKA, *University of Colorado*

78. Popular Fiction and Brain Science in the Late Nineteenth Century
ANNE STILES, *Saint Louis University*